Dutch Barns of New York
An Introduction

Dutch Barns of New York
An Introduction

by Vincent J. Schaefer

Preface by Thomas Lanni

PURPLE MOUNTAIN PRESS
Fleischmanns, New York

First Edition 1994

Published by
Purple Mountain Press, Ltd.
Main Street, P.O. Box E3
Fleischmanns, New York 12430-0378

Printed and bound in the United States of America.

All drawings and photographs, unless otherwise noted, are by the author.

Cover and title page: The Teller-Shermerhorn Barn, Schenectady.

Library of Congress Cataloging-in-Publication Data

Schaefer, Vincent J.
Dutch barns of New York : an introduction / by Vincent J. Schaefer : preface by Thomas Lanni. -- 1st ed.
p. cm.
Includes bibliographical references.
ISBN 0-935796-50-9 (pbk.)
1. Barns--New York (State) 2. Architecture, Colonial--New York (State) 3. Architecture, Dutch--New York (State) I. Title.
NA8230.S33 1994
728'.922'09747--dc20 94-12435
CIP

To the master builders of northern Europe who brought their vision and know-how to America and built their beautiful barns along the Hudson, Mohawk, and Schoharie valleys.

Table of Contents

Acknowledgments

I am indebted to friends in the Dutch Barn Preservation Society who have provided friendly and enthusiastic commentary on the content of this book, informing and fortifying my earlier and lonely conclusions about these unique barns of New York State.

Among many others, Clarke Blair, Everett and Peggy Rau, Mark Hesler, Chris Albright, Shirley Dunn, Greg Huber, Peter Sinclair, Clair Tholl, Sam and Elsie Van Orden, Thomas Lanni, Nanno Bienstock, Bob and Amelia Anderson, Carolina and Lee Lazzari, Skip Barsheid, Carl Touhey, Richard Babcock, Lou Caputzal, and John Kaufman in one way or another provided the inspiration for this book.

Special recognition is reserved for my dear wife Lois, who sustained me for more than 57 years and whose memory comforts me through these years remaining.

Preface

Five years ago, I attended a lecture at the Atmospheric Sciences Research Center by Vincent Schaefer, its Director Emeritus. To my surprise, he had chosen to speak not about the atmosphere or environment, but about an eighteenth century architectural form peculiar to the Hudson and Mohawk River valleys, the Dutch barn.

After the talk, I approached Dr. Schaefer. Though I am not a history buff, his words had evinced in me certain stirrings of interest, intrigue, and wonder. I happened to be living on an eighteenth century Dutch farm located on the Onesquethaw Creek, a tributary of the Hudson River in southern Albany County. I related this information, and it seemed to delight him, although I explained that none of our outbuildings seemed to conform to his descriptions. Beaming from ear to ear, he reiterated that Dutch barns were somewhat rare and, sadly, those that remained were disappearing from the landscape through disinterest and neglect.

It was only later in our acquaintance that I came to understand why Vince had been smiling as he told of the plight of these once great barns. He foresaw our meeting as serendipity, one of those accidental or unintentional events that brings about an insight or discovery of some kind. An important feature of such phenomena is that it is the open and prepared mind that most often enjoys them.

After that first meeting, whenever I met with Vincent Schaefer, I felt a certain excitement and sensed that I was involved in some greater purpose. Thus inspired, I proceeded to make inquiries in the neighborhood and at the library, and I began to learn much about the history of the area and of the farm on which I lived.

I learned that there had been a Dutch barn on our farm that its previous owner had misguidedly bulldozed in the 1950s. When I told Vince of my findings, he immediately suggested that we find another barn that needed a good home and move it to our farm. Within a short time, he convinced us to buy and restore the endangered Larger Wemp barn, one of the finest Dutch barns still in existence. The barn now stands proudly on the flats of the Onesquethaw. It serves as an educational museum and gathering place for school groups and the public, and is managed by the Dutch Barn Preservation Society.

I met with Vincent Schaefer many times, and, whether we talked of the environment, barns, or history, I always felt fortunate to have access to his vast store of knowledge. Perhaps the greatest legacy of my acquaintance with Vince is the heightened sense of curiosity he imparted to me. Like Vince, I now often take an unknown road in the hopes that a discovery may be made, as indeed some have. Now you, too, may share in Vince's experience and ruminations about this fascinating part of our local history and culture, the Dutch barns of New York.

—Thomas Lanni

The mid-eighteenth-century Decker-Bienstock barn, near Wallkill, New York.

Foreword

I have been having a love affair with Dutch barns for more than seventy years!

When I was about fourteen years old and living in the suburbs of Schenectady, I used to take Sunday afternoon walks with my father. We often went down the Tramp Path from Bellevue and crossed the Great Flats to the Old Erie Canal. This was my introduction to a region I've enjoyed ever since. Often on Saturday, while my father was working, I would pack a lunch and go off on my own, searching for arrowheads, violets, bull frogs, or wild strawberries, depending on the season. Around noontime, I frequently stopped to eat on a beautiful hillside near a large cold spring. I was so enamored with this spot that I resolved that when I grew up I would build my home on it. Fifteen years later, with my brothers, Paul and Carl, my uncle, Joe Holtslag, and my wife, Lois, I did just that.

Close to my choice spot was a huge barn, which I later learned was a Dutch barn. As I explored further, I began to notice other great barns marking the early homesteads of the Mohawk Valley. I soon learned to recognize these unusual structures from a distance by the austere simplicity of their profiles and the steepness and extension of their roofs.

When I was about sixteen years, old my father gave me his Kodak box camera and I soon became an avid photographer. A few years later, I acquired a folding camera, and started to photograph Dutch barns, an obsession still with me after more than sixty years! Back then, ten of the fourteen Dutch barns I knew about in Schenectady County were still standing. Now only one, the Wemple-Dellamont barn in the Town of Rotterdam, remains.

After photographing the Dutch barns in Schenectady County, I ventured further to discover others, notably the two Wemp barns, sitting side by side along the Mohawk River near Fort Hunter, close to the mouth of the Schoharie Creek. I photographed the Larger Wemp barn while it was still in use on a working farm.

I also visited the Van Bergen barn at Old Kaatskill, located near Leeds, New York, along the stream called the Cats Kill, and dating from 1680, the oldest known Dutch barn in the area.

Over the past sixty years, I have studied and photographed nearly a hundred barns built between 1680 and 1800.

In this book I describe and illustrate with sketches and photographs ten of these barns. Three of these now exist only in photographs, having been victims of fire, removal, or neglect. Two are only partially preserved and need repairs, four are preserved in private hands, and one is a brand new reconstruction following the design of one of the finest originals.

This book also includes material I prepared for *The Dutch Barn Research Miscellany*, a publication of the Dutch Barn Preservation Society, describing these buildings' construction.

In writing this book I hope to convey to the reader some of my fascination with these distinctive Dutch barns. They link us in a very real way to the earliest settlements in America by European pioneers.

Dutch Barns in America

The Dutch barns of northeastern America are unique architectural structures, providing strong cultural links to the countries of northern Europe, including Scandinavia and northern Germany, and particularly Holland. These substantial buildings were erected in the Hudson, Mohawk, and Schoharie valleys of New York, on Long Island, and in northern New Jersey—all areas of Dutch settlement.

The first of these barns were built fairly close to the edges of the rivers; but by the turn of the eighteenth century, farmsteads began to appear along secondary streams such as the Catskill, the Normanskill, the Roeloff Jansen Kill, the Wallkill, the Esopus, and the Onesquethaw, and similar streams flowing through rich, fertile flood plains and flats resulting from post-glacial Lake Albany about 10,000 years ago. The barns were located fairly close to the farmhouse sites and almost always near running water. Documentary evidence tells us that some few of these structures were used also as dwellings, as was frequently the case in Europe.

It is likely that barns had first priority in construction when a settler acquired a piece of land. The first house constructed was generally tiny and minimally sufficient for occupancy for the farmer's first few years of getting established. As the farmer prospered and his family grew (as many as six to a dozen children were common), a larger, more substantial residence was added onto the original structure, often in several stages.

The first barns were massive structures resembling the tithe barns of Europe. These barns embraced the same structural principles as the great cathedrals that were built around 1200 by northern European craftsmen.

The Dutch builders who arrived in the Hudson Valley must have prized the tremendous pine and oak trees growing in the surrounding forests and recognized the importance of proper seasoning before assembling the main timbers with mortise and tenon joints. Some of these builders were probably shipwrights, since many of the joints they utilized resembled those used in the construction of wooden sailing ships.

While the large barn timbers from the surrounding forests were being shaped and seasoned, simple hay barracks like those still in use in parts of Europe and Canada were constructed to protect the hay and grain from the weather. Some of these had elevated floors and thus provided some degree of shelter for livestock from rain and winter sleet and snow.

In 1674, the English wrested final control of New York from the Dutch; but the very conservative Dutch inhabitants of the farms and villages of the region continued to perpetuate their Dutch culture for more than a hundred years after the takeover. This innate conservatism continued to con-

Left. The Van Slyck Dutch barn (behind silo), near where the Plotter Kill enters the Mohawk River. This barn was destroyed by fire around 1960.

trol the design and construction of their farm buildings. The Dutch master builders and their apprentices duplicated the basic patterns for barn construction they were accustomed to following. Their barns built in the seventeenth and eighteenth centuries embodied elements of design already in use in Europe for several centuries.

In my estimation, the date of about 1760 marks a change in building style. However, the major dividing point between old and new styles of barns built to the west of Albany was the Revolution. In 178,0 Sir John Johnson and his band of a thousand British Regulars, Tory neighbors, and Iroquois Indians swept through the Schoharie and Mohawk valleys and murdering the Patriot settlers of the region and burning crops, houses, and barns.

The Dutch barns of the region that were erected to replace those that had been burned bear evidence that they were built hurriedly. I have a feeling that their workmanship does not compare favorably with their earlier counterparts, since the barns from after 1780 were built on an emergency basis. It is likely that the many innovative features that typify some later barns are the result of Yankee ingenuity rather than a tradition rooted in the Netherlands. There was the need to provide a sheltered crop storage space as quickly as possible. Although the barns' core structures remained the same, there was no time to include the niceties of earlier periods.

Residence of Fletcher Van Wie, Town of Root, Montgomery County. From History of Montgomery and Fulton Counties, *1878 (F. W. Beers & Co., New York).*

Origin, Structure, and Function

The features of the Dutch barn that distinguished it from all others were its roof profile, identifiable even from a distance, its floor plan, which tended to be square while the English barn was rectangular, and the support of its roof by a central core structure. Entry was from the gable end. Dutch barns ranged from 35 to 50 feet long. Their floors were built on stone piers a foot or more above ground level. The buildings were 36 to 40 feet high, with steep roofs sloped at angles of 79 to 90 degrees.*

All the weight of the roof rafters, sheathing, shingles, and any snow load was borne by the purlin plates and anchor beam posts, the floor, and the stone piers below. Little of the weight was supported by the outer curtain walls. The deep anchor beams held this simple but highly effective supporting structure together.

These beams of pine or oak stretched across the central threshing floor for a distance of 28 to 34 feet, often with tongues protruding a foot beyond the back of the anchor beam posts. On top of these posts was a purlin plate, often of a single timber running the full 50 or more feet from gable end to gable end. On top of this purlin plate were tapered rafters ranging in length (depending on the pitch of the roof and the height of the eaves) from 32 to 40 feet. On top of these rafters in the early days was often a plank roof whose overlapping planks were beveled to shed water.

This combination of floor, post, anchor beam, purlin plate, rafter, and plank roof formed the highly stable inner core, the earmark of all Dutch barns, and was responsible for the longevity and integrity of these barns. Because their outer walls were curtain walls, like those found on most modern skyscrapers, they could be removed temporarily without affecting the stability of the roof and the barn structure. Thus, the deterioration of the outer sills of a Dutch barn is of no immediate consequence except to the outer wall. In the later so-called English or German barns, such deterioration led to disaster, since their roofs and internal timbers were supported entirely by outer walls. Protected from moisture-attracted fungi, Dutch barns can easily last for two or three centuries and even twice that, and the beautiful joinery of their massive timbers is a joy to behold and attests to the skill of the artisan.

All true Dutch barns in America have this core structure, although many local variations occur. In hilly terrain, later barns were sometimes built into a hillside and an upper floor was laid on top of the anchor beam or on a recessed shelf cut into the beam's facing sides. This upper floor was reached from the side of the barn rather than by the gable. The side wagon doors supported by wooden hinges provided a drive-through mode of access.

Inside a Dutch barn, the soaring rafters rising above the massive anchor beams remind one of a cathedral. Indeed, among architectural historians, the floor plan of these barns is called *basilican*. The threshing floor between the anchor beams represents the *nave*, while the animal aisles on either side represent the *galleries* of the basilica, whose

* Roof angles refer to the angles included at the peak between rafters.

predecessors are found in the European cathedrals dating back to 1200 A.D.

The Dutch barn of northeastern America evolved from centuries of collective experience with this basilican design and of study of the construction of seaworthy ships. It is evident that these barns were built following pre-existing concepts and that they emerged in America from a full-blown design not needing any experimentation.

In the biography of General Ely S. Parker, chief aide of General U.S. Grant during the Civil War, there is an amusing commentary on the space contained within a Dutch barn. A huge building, erected to exhibit American manufactured equipment and handicrafts at a national fair in Washington after the war, reminded one observer of a Dutchman's barn it seemed to have more room inside than out!

The space within did seem to be larger than anticipated from the barn's exterior appearance. When in active use, the structure was a scene of intense activity throughout the year. It is sad to visit an abandoned Dutch barn and find it cluttered with discarded farm implements, household debris, and the cast-offs of a computer-driven economy. Stripped of clutter, such a barn, if it possesses a good roof, could offer itself to a variety of uses limited only by the imagination of its owner.

The Dutch barn's primary function seems to have been to provide for the protection, storing, and threshing of wheat and other grains. The threshing floor and the granary were integral and important parts of the barn's interior. The "martin holes" and wagon doors at both gable ends provided adequate ventilation and encouraged birds, particularly purple martins, to feast on the insects brought in with the harvest. On a central anchor beam, an assembly provided support for a vertical post that extended from a socket in the under surface of the anchor beam to a hole below it in the threshing floor. This post had a special harness to which a horse or a mule was fastened. The animal was then driven around a circular path so the pressure of its hooves would release the seeds from the tassel.

My informants on early farming practices, Hugh (Sam) and Elsie Van Orden, described how contamination of the grain during the threshing was prevented. Before being put into harness the animal was tethered outdoors to have his underside sprayed with cold water, and a bag was hung between his rear legs to catch solid droppings.

Both before the barn was built and later, it is likely that some of the harvest was stored in a hay barrack similar to those still used in parts of Europe and Canada. Outside storage protected a barn from fire if spontaneous combustion ignited the center of the hay stack. A hay barrack consisted of a movable roof that rode up and down on four to six 25-foot poles projecting through holes at the edge of the roof. Holes placed about 12 inches apart on the vertical uprights held horizontal rods that supported the roof at a level slightly above the top of the piled hay, straw, or shocks of grain. Sometimes the bottom of the hay or grain stack was elevated 3 to 6 feet above the ground to provide protection for farm animals and domestic fowl.

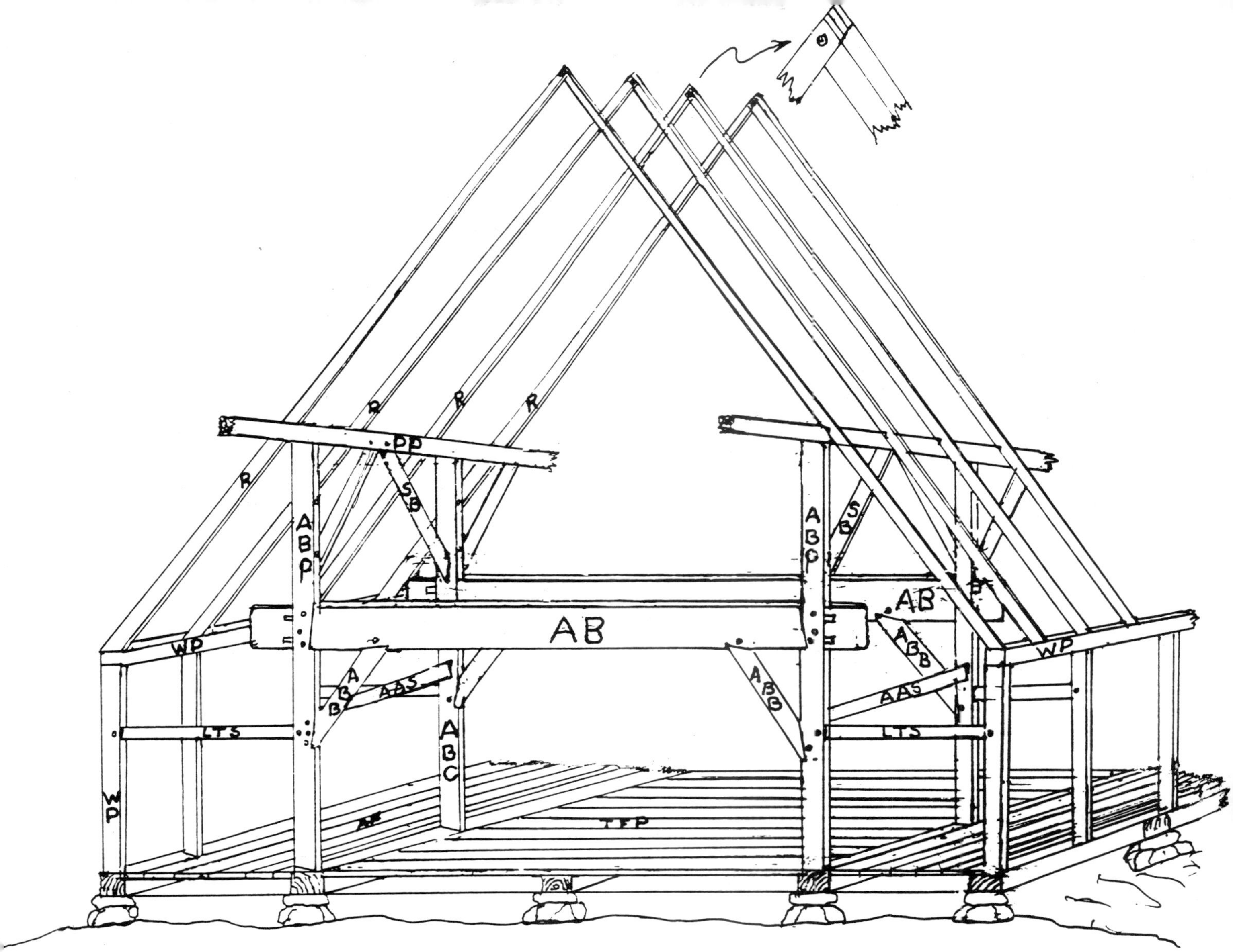

AB	*Anchor Beam*
PP	*Purlin Plate*
ABB	*Anchor Beam Brace*
ABP	*Anchor Beam Post*
WP	*Wall Plate*
TFP	*Threshing Floor Planking*
AF	*Aisle Floor*
SB	*Sway Braces*
LTS	*Lower Transverse Struts*
WP	*Wall Post*
R	*Rafters*
AAS	*Animal Aisle Struts*

A pair of interior "bents."

Three or four bents joined together make up a two- or three-bay Dutch barn. The anchor beams at the gable ends often do not have braces. The central core "H-frame" (the anchor beams, anchor beam posts, and purlin plate) supports the roof. The outer walls are non-supporting "curtain walls." Sections outside the H-frame are called aisles. It is this core structure, flanked by aisles, that distinguishes the Dutch barn from all others.

The Teller-Schermerhorn barn reveals its structure during dismantling by the author in 1947--1948. Above. View from the south. Below. View from the southwest.

Roof Angles

The Dutch barn roof had a steeper slope than that of later barns. The oldest barns had the steepest roofs, sloping at an angle of around 80 degrees. After about 1760, as side walls became higher and roofs became flatter, the slope approached 100 degrees. Barns built after the Revolution to replace those burned or otherwise had roofs sloping at angles of 105 to 110 degrees.

My Amsterdam, Holland, correspondent, Jaap Schipper, suggested in his letter to me of September 1989 that it is likely the builders of the seventeenth century figured the slope of the roofs of their barns using a fixed ratio of so many feet of the barn's height against so many feet of its width. In this manner, the slope of the Teller-Schermerhorn Dutch barn, which I had to dismantle in 1947-48, was constructed according to a 1:1 ratio, a classic ratio duplicating that used on many barns in Holland. Half of the western roof had a lesser slope of 5:7. This was similar to examples of roofs of house/barn combinations built in Holland in 1548 and 1650, respectively

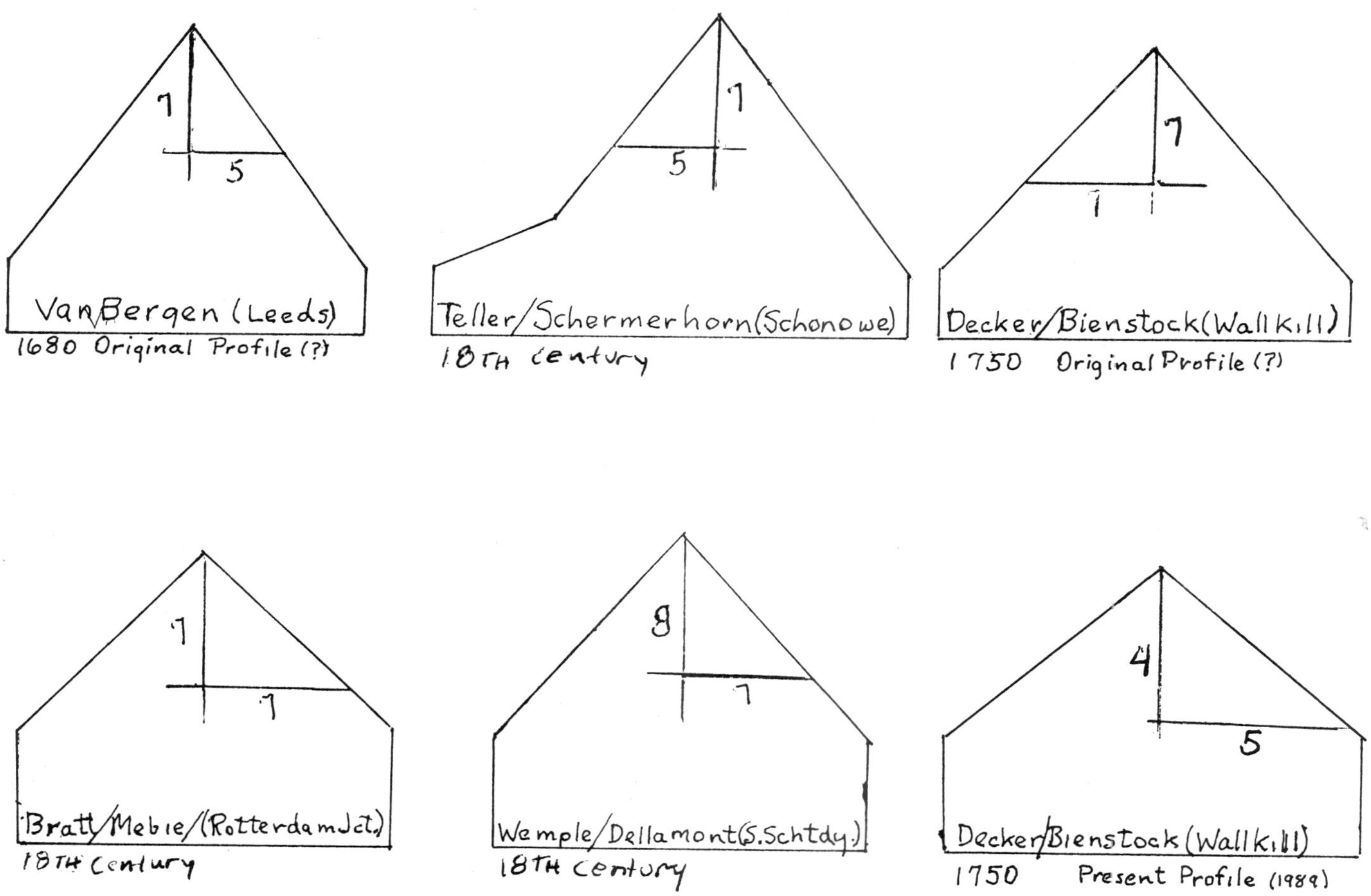

Foundations

The foundations of Dutch barns consisted of piles of flat or semi-flat stones arranged strategically to support the sills of the barn around the periphery and under the posts supporting the anchor beams. These stone piers were generally of native sandstone or limestone that had split naturally into convenient thicknesses to provide excellent stability. They enabled the floor of the barn to be suspended a few feet above the ground so as to permit adequate air circulation beneath the barn floor. Since the ground beneath the barns was dry, frost heaving was no problem. In some instances, if the barn was located on a moderate slope, a dry wall foundation was built so that the sills were level. This did not interfere with good ventilation.

The sills were massive, frequently 12 by 12 inches in cross section, and each was originally a single long piece of wood. In large barns, a median sill ran the full length of the barn from gable end to gable end, supporting floor planks for a bit less than half the width of the threshing floor.

Floors

The floors of many of the remaining Dutch barns are original. They consist of oak planks 3 to 4 inches thick, 12 to 15 inches wide, and 12 to 16 or more feet long, resting on top of the sills. In some instances, the planks are chamfered to fit into the median strip. In many instances, the knots in the wood of these original planks now stand out one-half inch to an inch higher than the surrounding softer wood and are often highly polished from two or more centuries of use.

Sandstone dry wall supported the Teller-Schermerhorn barn lean-to.

Other floors were made of split pine planks. The early planks were 6 inches thick and as much as 18 to 20 inches wide. Their upper surface was adzed flat, while underneath, notches were cut to fit onto the floor sills. This produced a floor thickness of 3 to 4 inches.

The Marten Van Bergen barn of Old Catskill had an oak floor made of trees shaped with parallel sides that were laid close to each other on the dirt. Their third surface was adzed so as to make a flat floor. The heavy floor of the Zaretski barn (ca. 1760) near Guilderland Center was made by recessing planks six inches thick so they would rest on the sills even with adjacent thinner planks. Such a method of floor plank preparation was necessary if there were no sawmill to carry logs of the size needed. Frequently planks were fastened to the sills with wooden pins similar to the oak pins used to secure other timbers. While oak planks appeared to be the wood favored for floors, white and pitch (yellow) pine were sometimes used. Pine was not as resistant to wear and rot as oak, but it had considerable durability as long as it remained dry.

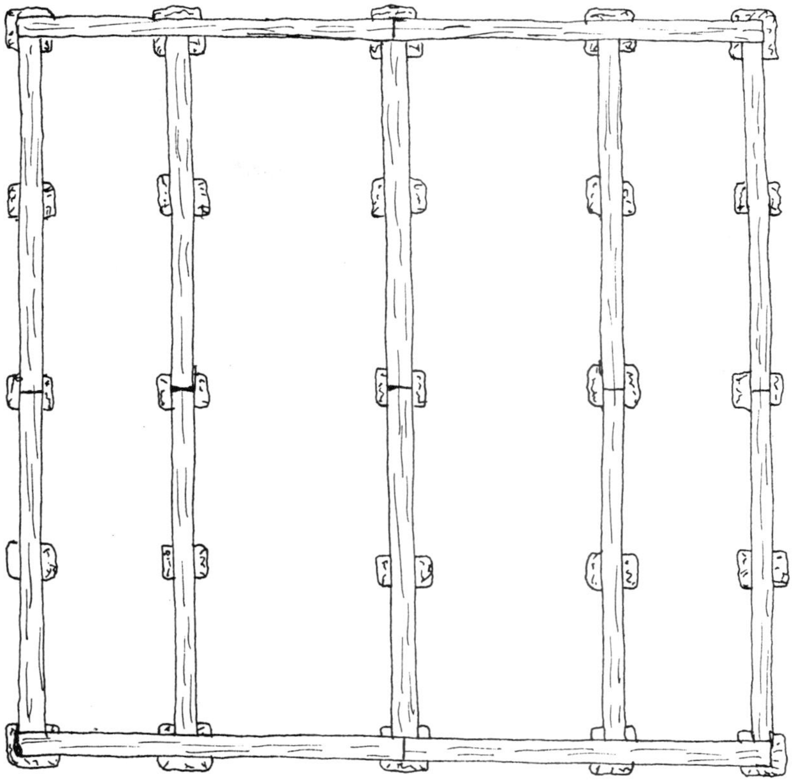

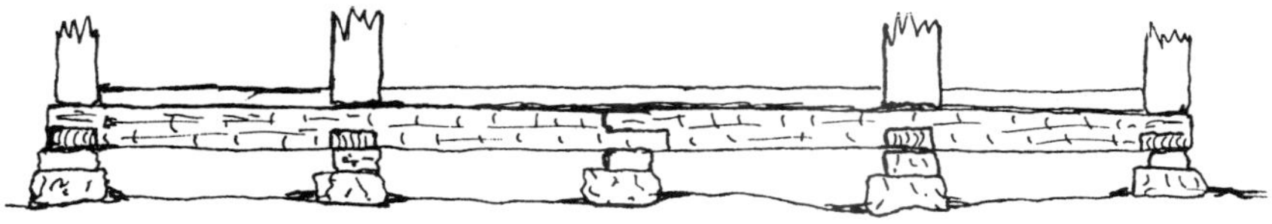

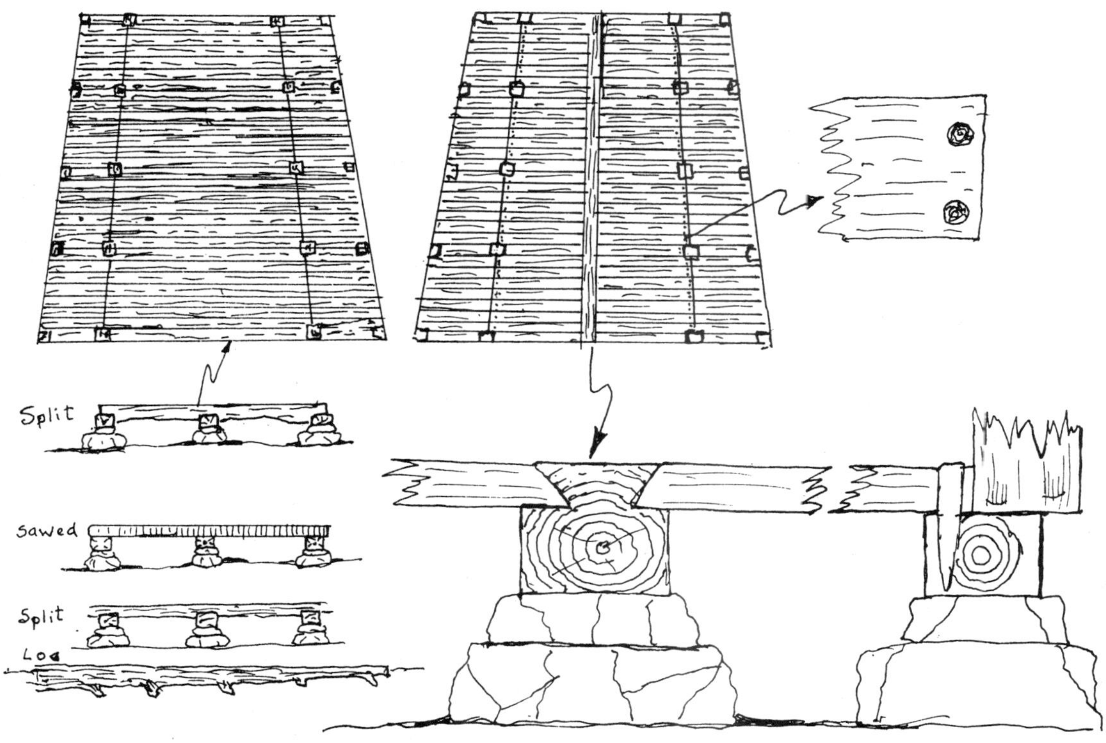
Split
Sawed
Split
Log

Anchor Beam Posts

The roof of a Dutch barn is supported entirely by the *purlin plate*. This is a long, fairly large (about 6 by 8 inches) support running the length of the barn from gable end to gable end. In earlier Dutch barns, the purlin plate might have been a single timber. The plate, in turn, is supported by large vertical posts joined by a massive horizontal timber, the *anchor beam*.

The vertical anchor beam posts delineated the bays in Dutch barns. The posts were generally of massive size, with cross sections of 10 to 14 inches that were either square or rectangular by one or two inches. They were frequently made of oak or pitch (yellow) pine and were tenoned into the purlin at the top. Depending on the height of the side walls, these posts extended five to ten feet above the top of the anchor beam they supported. A rectangular hole, the mortise, was cut into each post to fit the large anchor beam tenon. This mortise ranged in width from 3 to 4 inches and in depth from 16 to 23 1/2 inches. At the base of the mortise the post was recessed one to two inches so

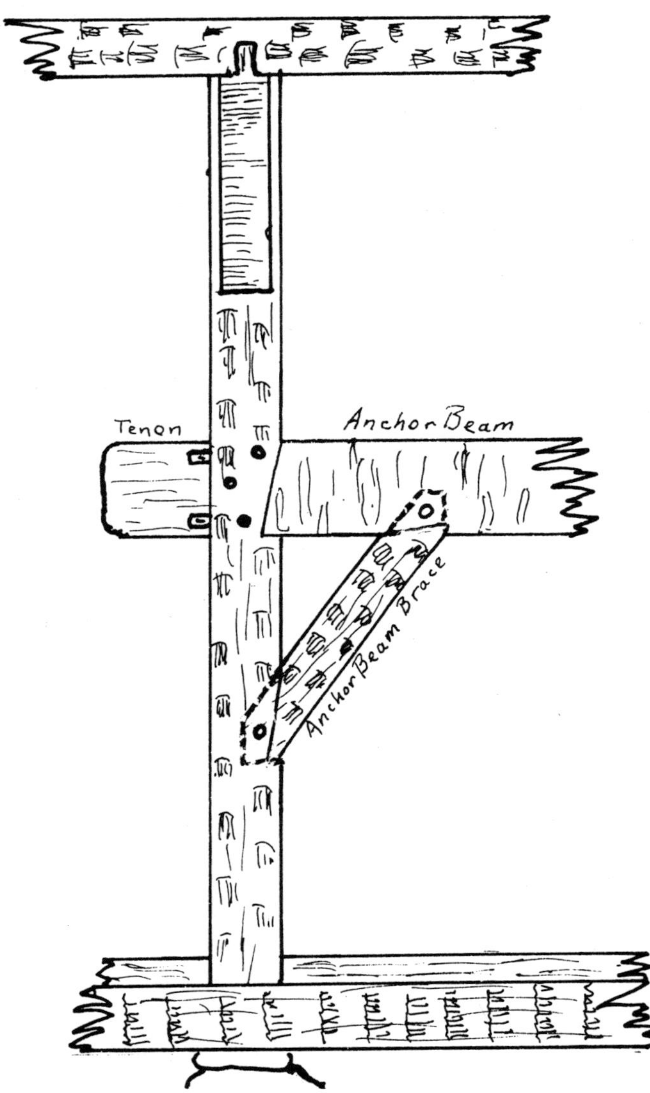

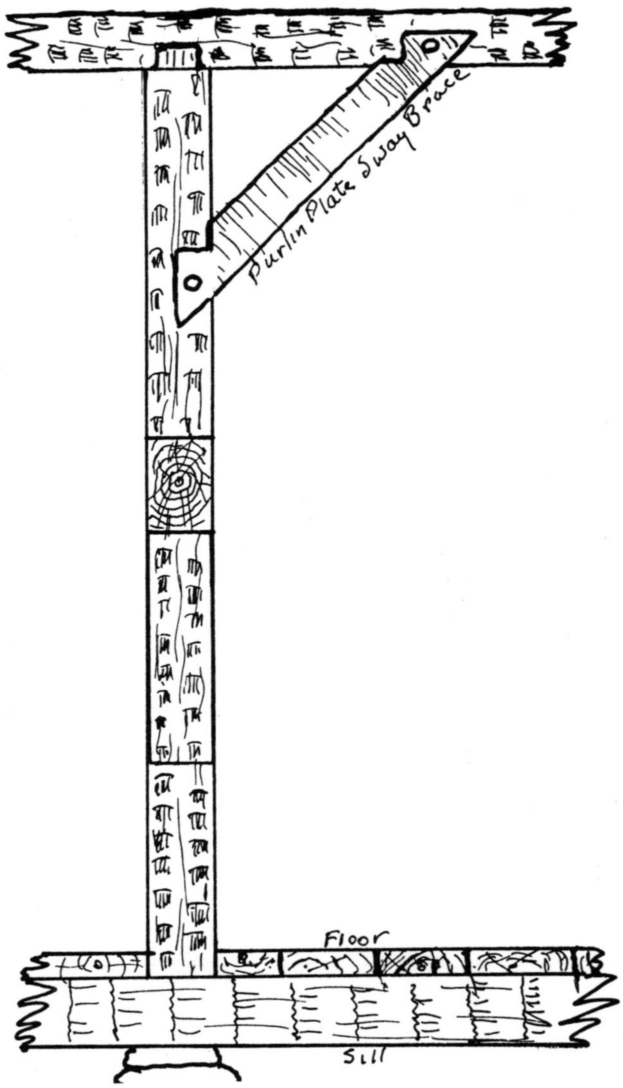

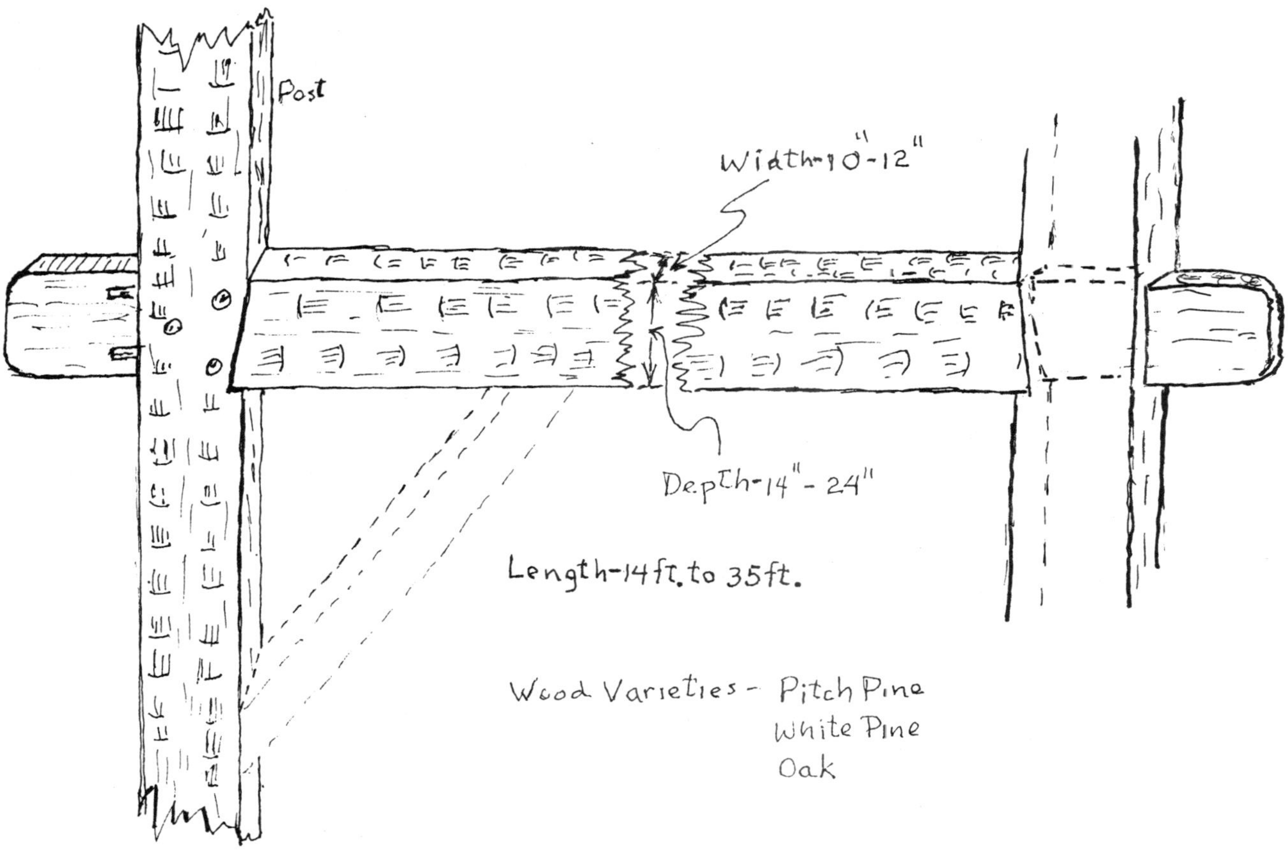

that the bottom of the anchor beam could rest there. Thus the weight of the anchor beam was supported by the shoulder of the post rather than depending solely on the strength of the tenon.

Anchor beams range in cross section from 8 by 10 to 12 by 22 inches. Generally, earlier barns had the more massive beams. Anchor beam braces were mortised into the sides of the posts and the undersides of the anchor beams at an angle to the posts of 35 to 45 degrees. Sway braces were mortised into the underside of the purlin plate and the nearby post. In earlier Dutch barns the brace was as wide as the anchor beam. In later barns it might have been half the width or even consisted of two smaller parallel braces.

These joined timbers making up a single bent were placed at 10 to 12 1/2 foot intervals, depending on the number of bays to the barn. There were usually three to seven such bays, the most common number being four.

Anchor Beams

These hewn members now known as anchor beams were the show pieces of the Dutch barn. They were usually made of white pine or pitch (yellow) pine. Over several centuries, those made of white pine developed a beautiful brownish patina. The earlier, more massive beams were roughed out with a broad axe and often finely finished with a very sharp adze. Some looked almost as though they had been finished in a planing mill. The patina of aging pitch pine was not as beautiful as that of the white pine, though the contrast between the dark pitch and the lighter spring wood was dramatic. The barns I have seen in which oak was used for the beams, in the mid-Hudson Valley and New Jersey, are all of a later vintage, dating from about 1760. Presumably, these regions did not possess the virgin pine forests of the more northern locales. It is likely that the Pine Plains extending from Albany to Schenectady, over the ancient, sandy, post-glacial bed of Lake Albany, were a ready source of massive virgin pitch and white pines.

Anchor beams ranged from 20 to 36 feet long. Three to four feet of this length served to form the massive tenons that projected through the posts and provided back-up protection against splitting once the assembly was completed. Two rectangular wedges made of oak or hickory held the anchor beams snug against the supporting posts.

Anchor Beam Tenons

A quick glimpse into a barn reveals it to be a Dutch barn if the anchor beam tenons project beyond the rear side of the massive vertical posts supporting the purlin plate that in turn carries the roof rafters. A few barns exist where the tenons are flush with the rear side of the posts. In some of these cases, the tenons have been cut off for no good reason; but in a few cases (such as the Deertz barn recently moved to Ancram, N.Y., from the Schoharie Valley near Middleburgh), the tenons were originally meant to be flush with the back surface of the post, as proven by their chamfered edges.

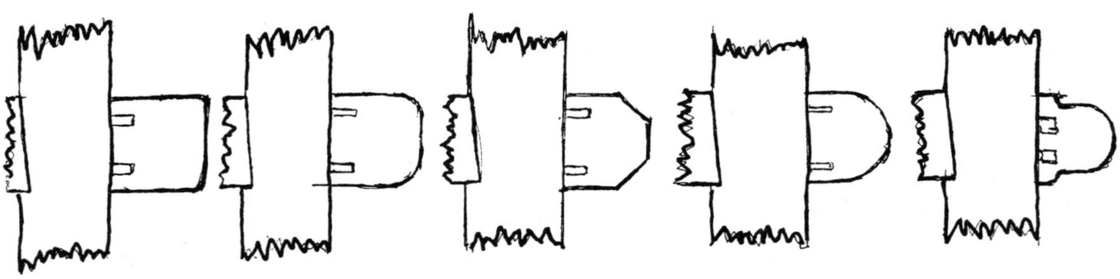

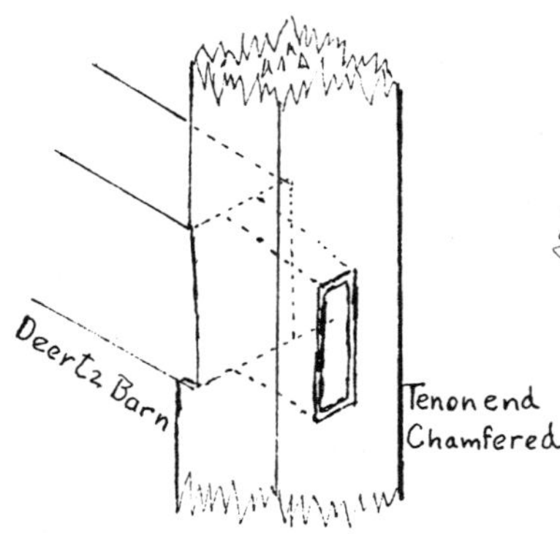

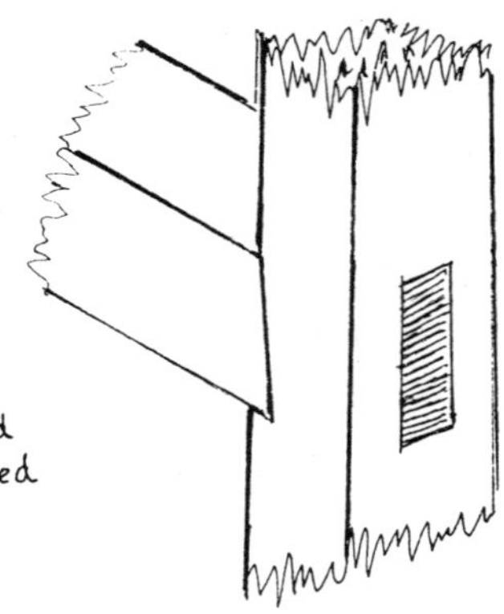

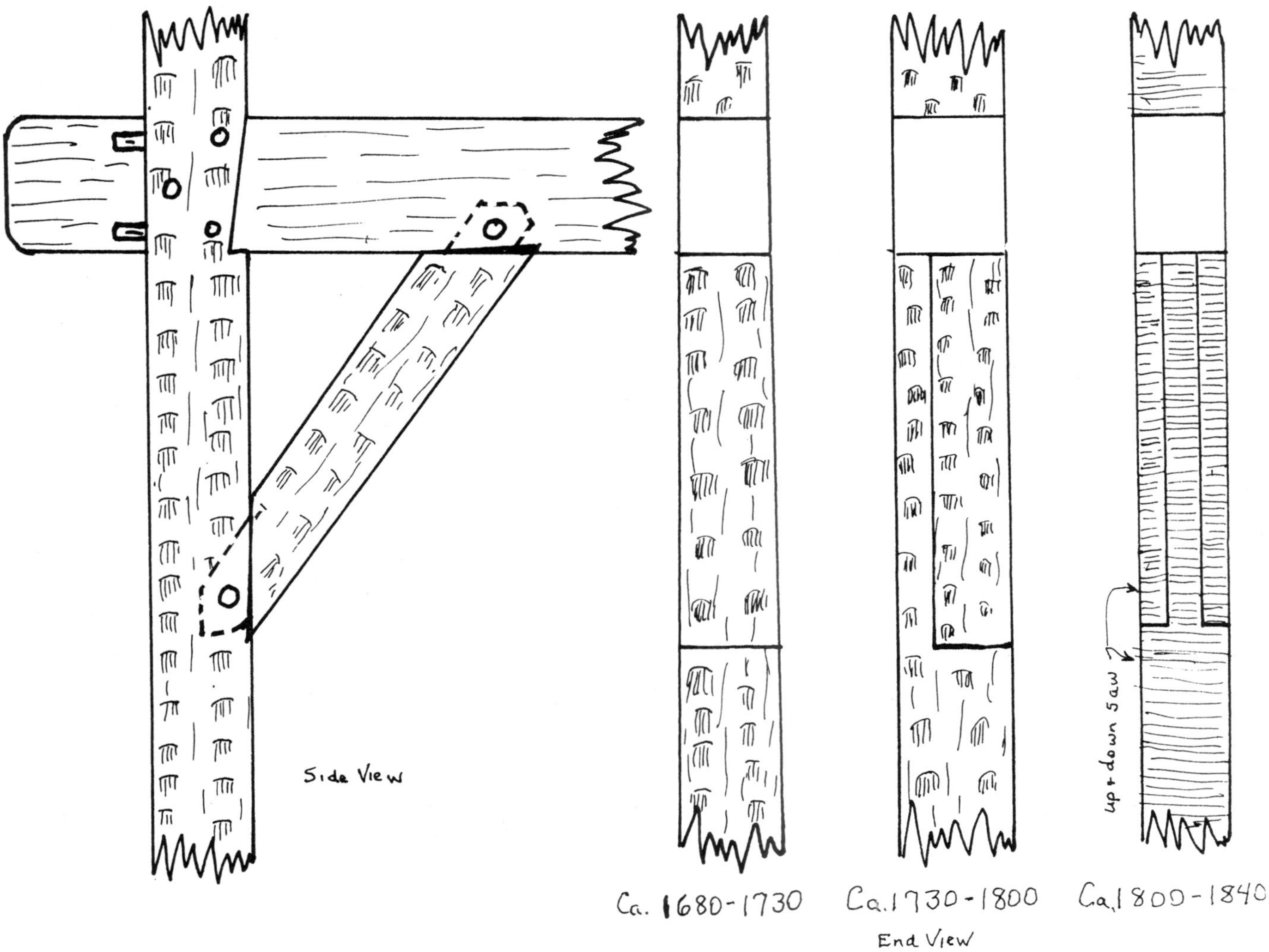

In most instances, the tenon is three to four inches wide, the full depth of the anchor beam, and projects from ten to twelve inches beyond the back surface of the supporting post. Some tenons are still square-ended, but the majority have their corners round or cut at 45 degrees. In most instances these tenons have been nicely made, fitting into mortises in the post with very little clearance. This joinery reveals the quality of the barn builder. When properly fashioned, these joints are a joy to behold. Even after more than two centuries, a well-built barn can have joints that are still in near-perfect condition.

Anchor Beam Braces

Part of the joinery of the anchor beam/post complex is the anchor beam brace. This is a timber that extends from the front surface of the post to the under surface of the anchor beam. The brace has tenons on both ends that fit into mortises cut into the upright and horizontal surfaces of the post and beam. The assembly of these three timbers is accomplished at ground level, before the frame is raised, by inserting the brace joints as the large anchor beam tenon is slid into place. Once seated and the wedges driven home, wooden pins are driven into holes cut into the post and the beam to make an assembly which, if kept dry, will last for centuries.

It appears that the dimensions of the anchor beam braces provide an important clue to the age of the barn in which they are found. As a general occurrence, the oldest barns have braces the full width of the anchor beam and the post which holds it. In later years, the braces sometimes consisted of two pieces of wood, their combined widths still less than the width of the beam or the posts making up the complex of the H-frame.

In the oldest barn of which we have firm measurements (the Marten Van Bergen barn of Old Catskill, 1680) the braces were the width of the post and anchor beam and the soffit of the brace was curved. In this barn, there was a second set of cross members attached to posts at an angle of about 30 degrees from the vertical, and a second set of purlins, making the framing similar to that found in houses in the Netherlands.

Purlin Plates

The purlin plate is the long horizontal timber running the length of the barn and resting on top of the anchor beam posts. The purlin plate was made of a single timber sometimes as long as sixty feet with a cross section of 10 by 12 inches or 8 by 10 inches. Sway braces were pinned into mortises cut into the under surface of the purlin plate and the sides of the anchor beam posts.

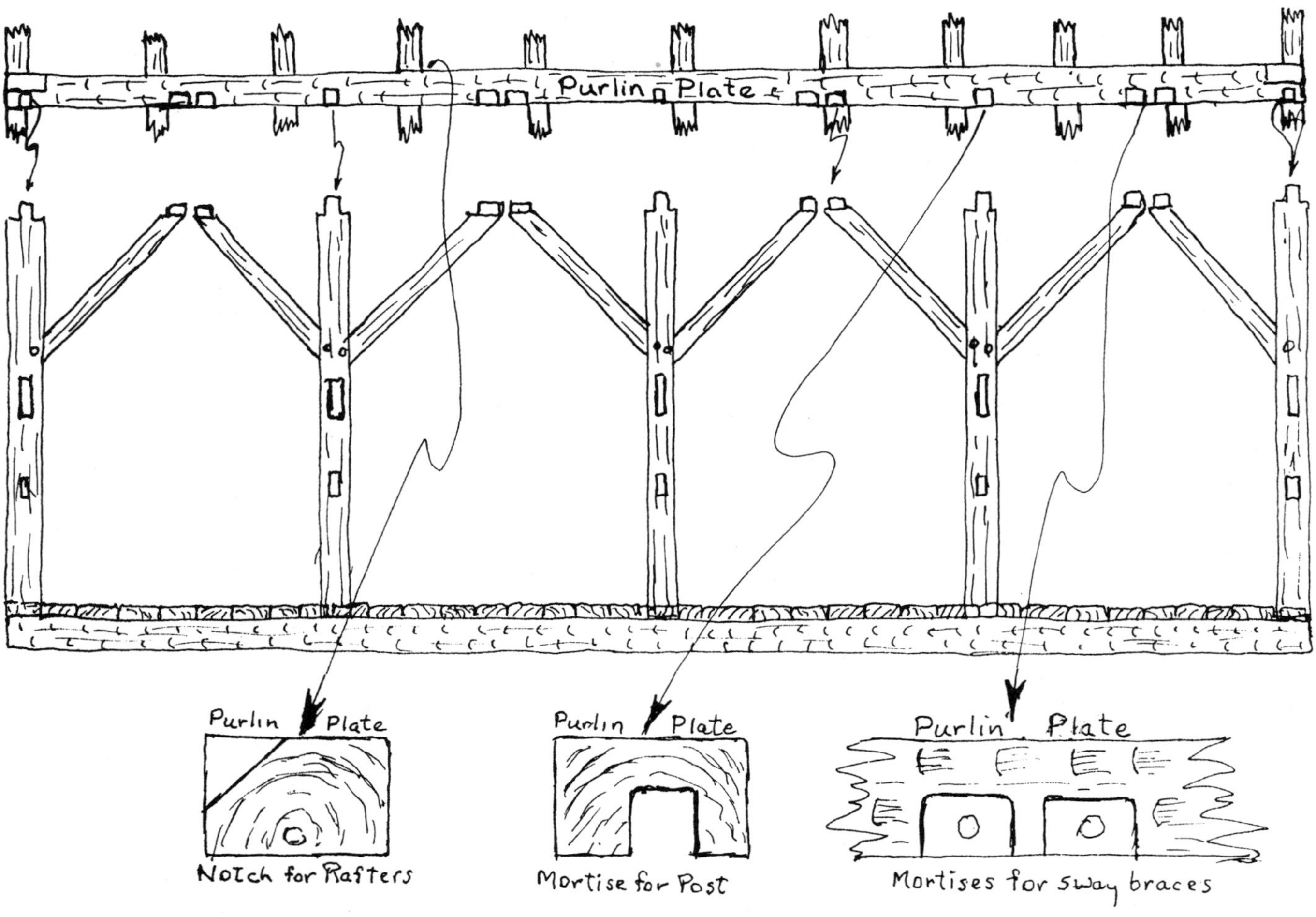

Purlin Plate Sway Braces

As support against the wind load on the roof of the Dutch barn, a series of timbers attached from either side of the anchor beam posts to the underside of the purlin plate was employed. These sway braces were mortised and tenoned and inserted when the purlin plate was raised.

The earliest barns had the longest sway braces. A sway brace frequently extended from the underside of the purlin plate near an adjacent anchor beam post down to an opposite post where it was inserted into a mortise cut *below* the anchor beam. In a few instances, a long brace would be crossed by another and fitted to it at the crossing, so as to make of two fairly long sway braces to a bay.

More frequently, sway braces were considerably shorter, so that the upper ends of these braces met under the purlin plate at the center of the bay, while their lower ends were inserted into the sides of the anchor beam posts to make an angle of 45 degrees.

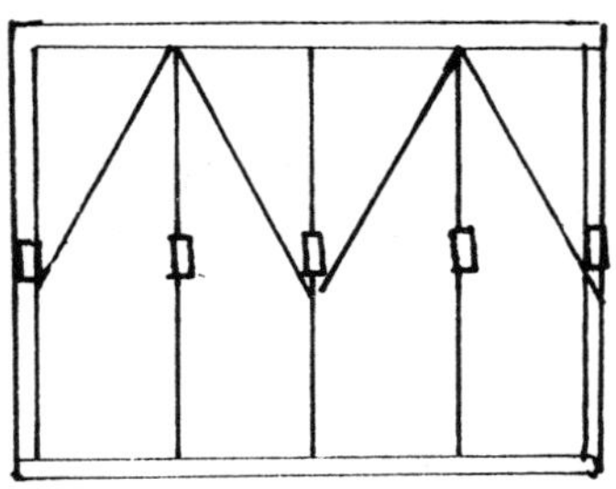

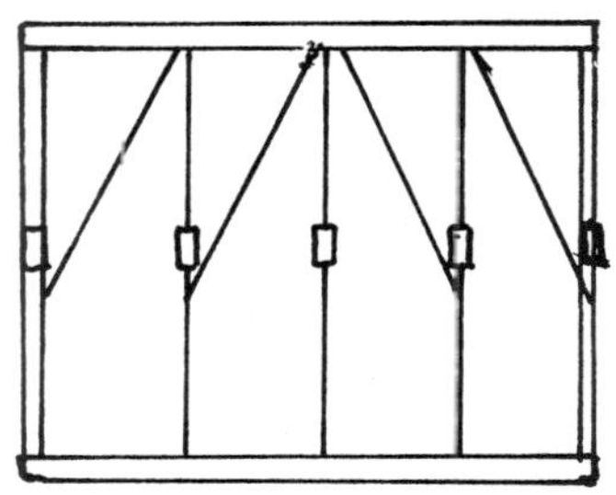

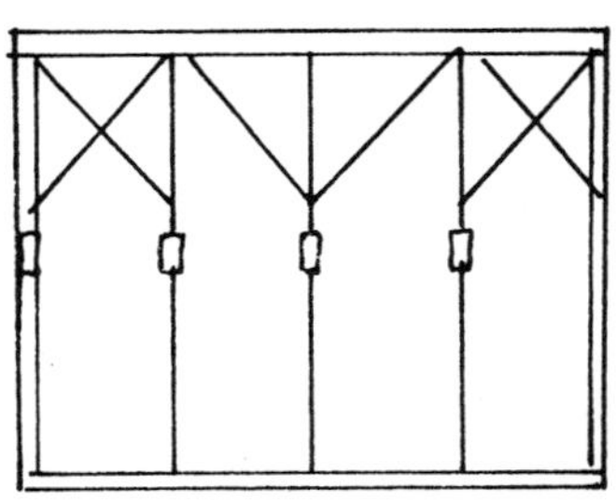

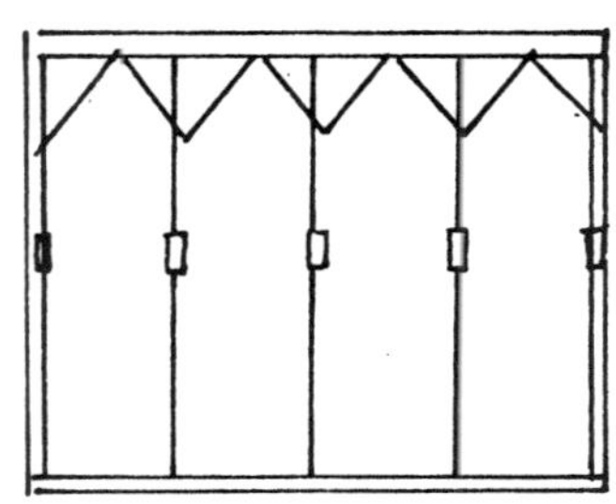

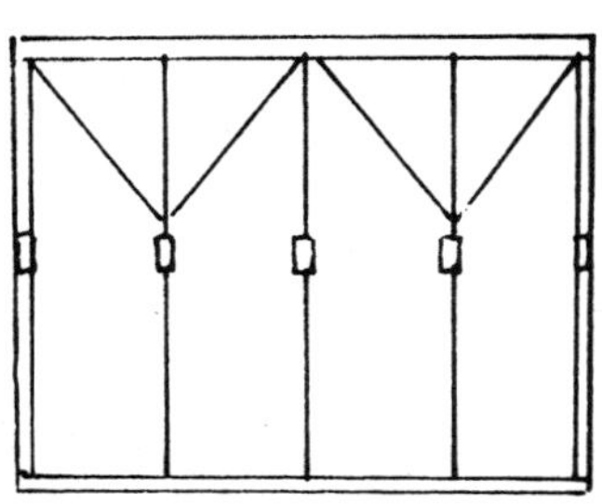

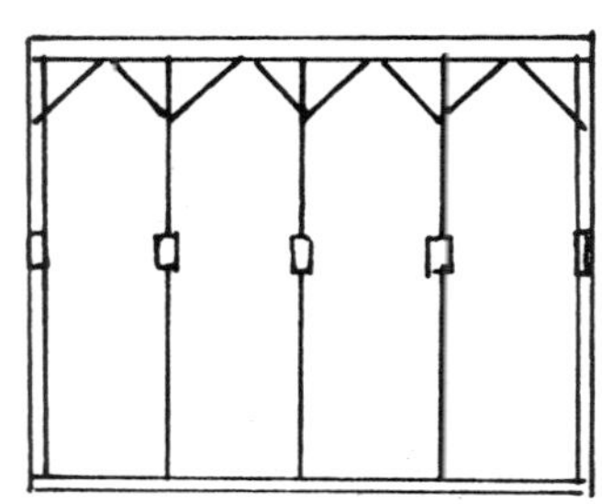

Above. Anchor beams and braces from the Teller-Shermerhorn barn.
Below. Interior of Van Bergen barn; sway braces mortised below anchor beams.

Above. Anchor beam brace with curved soffit from the Van Bergen barn.
Below. Massive anchor beam and brace from the Wemple-Dellamont barn.

Five rafters supported by the purlin plate in the Teller-Shermerhorn barn. Their bases originally rested on the wall plate, which has been taken away.

Roof Rafters

Especially in older Dutch barns, the roof rafters are big timbers. A barn commonly has 8 to 15 pairs of tapered rafters, with each rafter 32 to 40 feet long. The most common number of pairs of rafters is 11 in a 4-bay barn.

The rafters are tapered to range from 5 by 5 or 6 by 6 inches at the top to 7 by 7 or 8 by 8 inches at the bottom. The tapered nature of the timbers allows them to rest in perfect balance on the purlin plate. The rafters are hewn and occasionally smoothed by adze.

There is never a ridge pole in a Dutch barn. The upper ends of the rafter timbers are finished with a fork-and-tongue fitting, into which a wooden peg is inserted to hold the fitting together. Most often, the rafters are notched so as to set firmly on the purlin plate, to which they are then secured by a very large iron spike or a long pointed wooden peg. Frequently, the rafters are spaced about 40 inches apart, two of them sharing the space between the bents and a third located near the anchor beam posts. As the rafters are not directly above the posts they do not interfere with the mortise and tenon connecting the posts with the purlin.

There is evidence that some earlier barns were made with double-sloped roofs like those built in the 1600s in the Netherlands. The Teller-Schermerhorn barn had a fourth of its roof made this way. Three barns in the mid-Hudson Valley, near New Paltz---the Decker, Jansen, and Olive barns---have posts that were spliced so as to make this change possible as additional support against a load of snow. In each of these barns, the height of the posts was doubled in length above the anchor beam by the splice.

Roofs

The first roofs of the Dutch barns in our region were probably made of overlapping planks, although thatch was also used in the seventeenth century. The planks were 12 by 13 inches wide and fabricated with a pit saw or an up-and-down water-powered saw. Each side was given a wide chamfer and overlapped to prevent leaks; but due to imperfections in the lapping, the roofs leaked anyway, so they were eventually covered with white pine shingles. At first, these shingles were probably cut with a froe and installed despite their irregular surfaces. As they deteriorated, they were replaced by a variety of roof materials ranging from sawn cedar shingles to asphalt strips and metal.

In recent years, metal sheeting has been used for many barn roofs. When properly installed, metal roofs will probably last longer and are less expensive than any others.

In a few instances, slate roofs have been installed. When made of a good grade of material and properly installed, a slate roof has a nearly indefinite life span; but nowadays

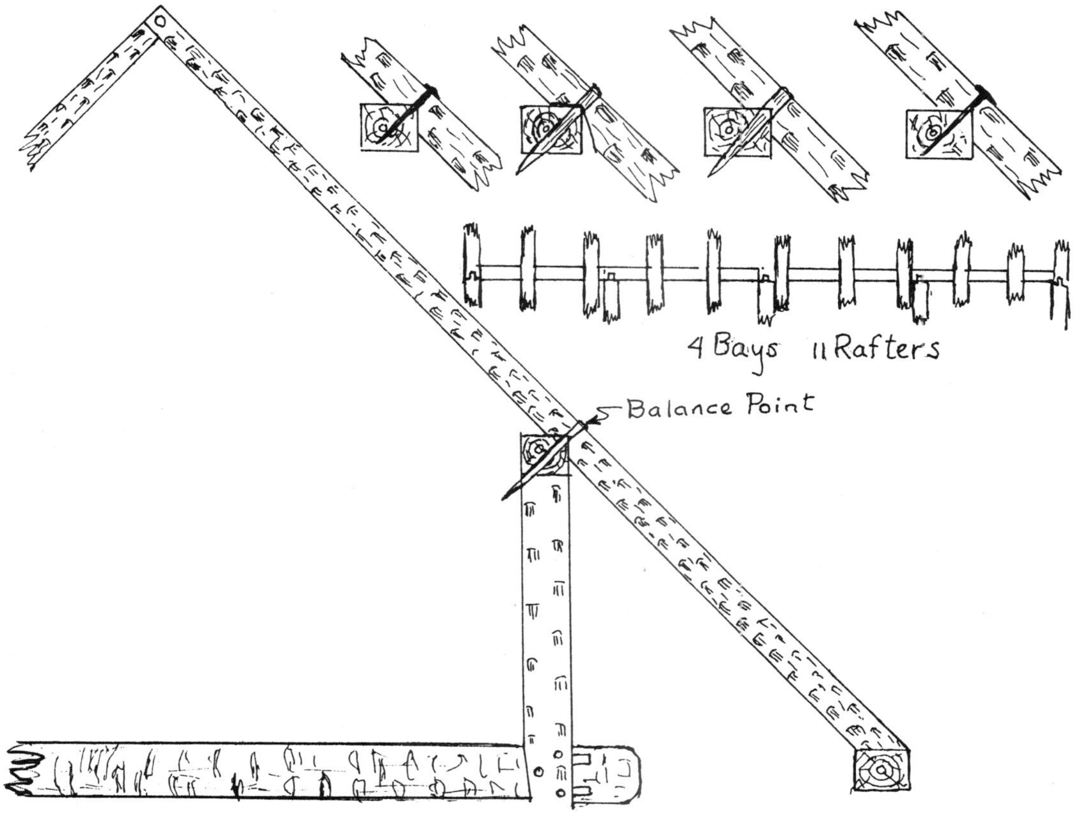
4 Bays 11 Rafters
Balance Point

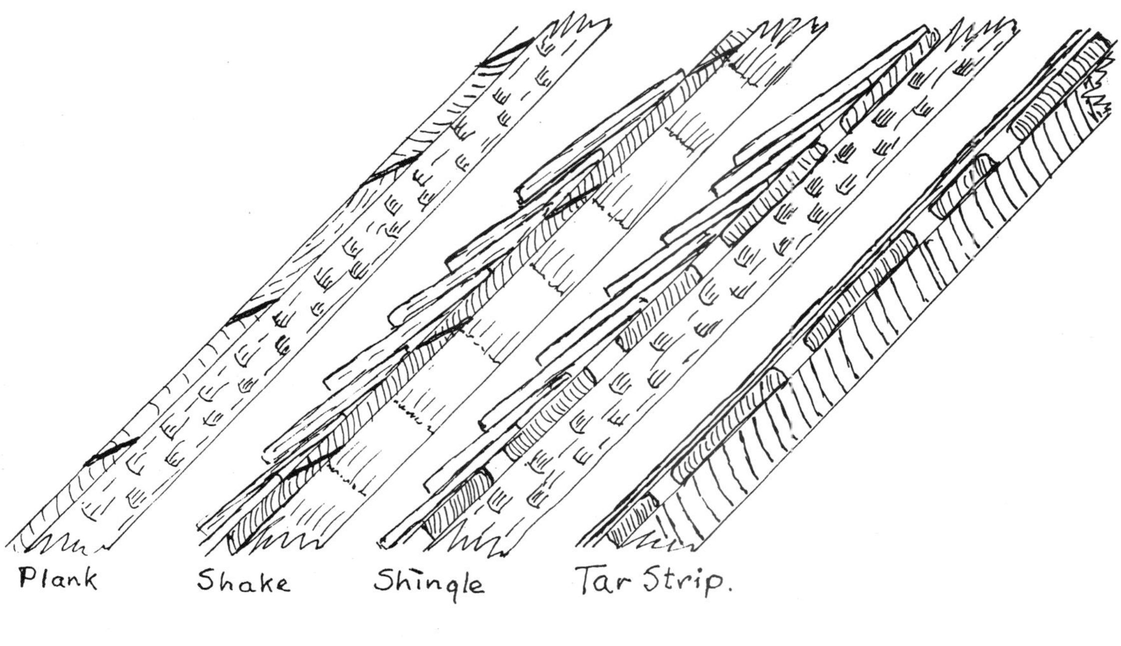
Plank
Shake
Shingle
Tar Strip.

they are excessively expensive. And a heavy slate roof may stress the barn frame.

Overlapped solid plank roofs evolved into shingled roofs with underlying boards that were butted and then spaced several inches apart. This treatment probably lengthened the life of the roof's surface material by providing better ventilation to its underside. The roof boards were at first quite wide, but, in later days, they were made narrower so that some were no more than 6 inches wide.

Earlier barns were fitted with wooden eaves troughs, intended to channel rain and melted snow away from the sills of the barn. These troughs were supported by pegs inserted at the top of the wall timbers or the wall posts. The eaves troughs were large enough so that in one instance (the Robert Andersen barn in Westerlo) a discarded eaves trough was later used as a horizontal timber in the flooring of the adjacent house.

Threshing Floors

The large floor space under the anchor beams was used to thresh wheat, oats, barley, rye, and other seed-bearing crops. Originally, the threshing was done with flails, wooden devices loosely fastened to handles which were used to hit the ripe head of the grain to release its seeds. After the seeds were flailed, the straw would be raked away and the loose seeds gathered, separated from the chaff, and then stored in the granary.

Later the threshing was done by the hooves of an animal harnessed to and driven around a pole (see description on page 16).

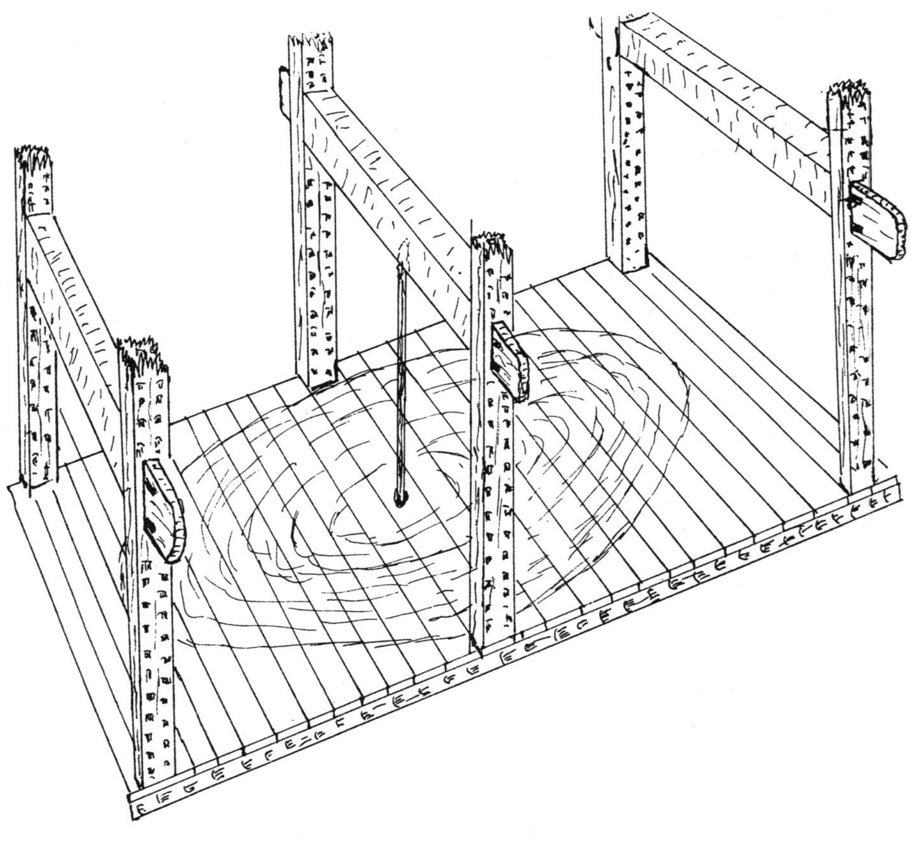

Animal Aisles

On either side of the threshing floor were aisles for horses, cattle, and other livestock. Often the granary was located at one end of a side aisle. This was a sealed room with a tight-fitting door of very wide tongue-and-groove boards and a solid hole-free floor and ceiling. This careful construction was designed to house the threshed grain and keep it rodent-proof. Still, rodents managed to gnaw holes in the siding and get to the grain, so metal sheets would be added to cover holes and prevent further depredation.

In a number of barns, a manger for dispensing hay to the animals was constructed along the threshing floor side of one side aisle. The upper part of the manger often was built with a series of parallel slats or round saplings fitted into the timbers extending along the aisle and fastened to the anchor beam posts. The slats slanted away from the side aisles in the direction of the threshing floor, since the heads of the animals faced the threshing floor. Mangers disappeared from most barns as wheat growers converted to the production of milk, and stanchions and other more modern devices were added to accommodate the increasing number of dairy cows.

Each side aisle had a small animal door at one or both gable ends of the barn. Some were single, others were double doors. Almost always they swung on wrought iron Dutch and strip hinges of various configurations, many of which were quite beautiful.

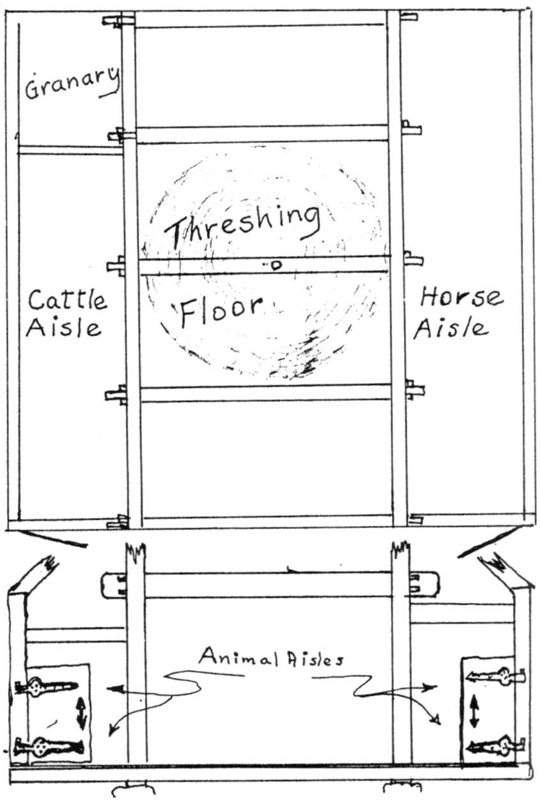

Right. A pair of wooden grain forks found in the Van Eps barn near Hoffman's Ferry.

Central Doors

Another feature setting apart Dutch barns from later English, German, Shaker, Pennsylvania, and other barns was the pair of large central doors located at each gable end of the barn. These were about 10 by 10 feet, consisting of a two or three components and swinging inward to accommodate a wagon loaded with hay or grain. Once unloaded, the wagon could be driven through the barn and out the doors at the other end without the horses having to back up.

Of the doors' components, one was usually full-length, the other split into upper and lower halves. The upper half could thus be opened for ventilation, while the lower half remained closed.

The doors were hung on wooden hinges that extended to nearly the full width of the door. They were secured by pins of oak or hickory. Sockets were cut in the door's framing posts to guide the hinges, which were retained by round dowels that fitted into a groove cut above and below the socket. Seven or eight rosehead nails held the hinge to the door. Divided Dutch doors had an extra two hinges. Early Dutch barn hinges were so well installed that, after more than 200 years, some still swing true.

In most instances, the old inward swinging doors have been replaced by doors that swing outward on iron strap hinges or slide parallel to the barn siding on iron tracks. In some later barns, the gable end doors have been replaced by side doors. Later barns had higher side walls of fourteen to sixteen feet, while earlier barns had walls from seven to ten feet high.

In the earliest barns, a "people" door often was located adjacent to the wagon doors, allowing entry to the barn for the farmer without his having to open the more ponderous wagon door. This door also hung on Dutch hinges and often had a wooden latch or bar to keep the door closed when not in use.

Right. Wagon and people doors in the Van Bergen barn as they were in 1957.
Next page. Wagon doors of the Haselo-Viele barn from the inside.

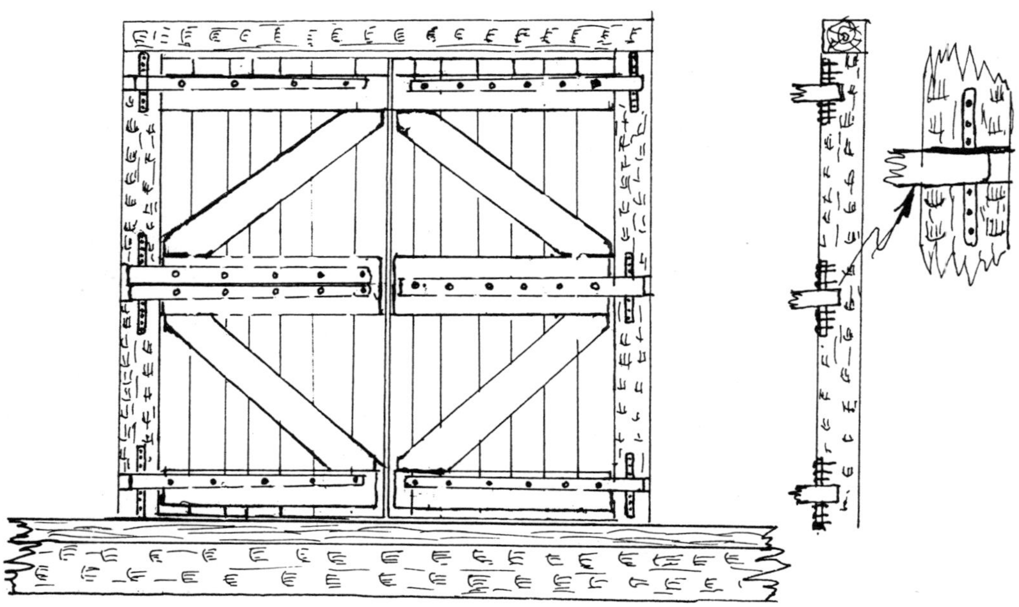

Animal Aisle Doors

At the extreme corners of one of the barn's gable ends were either single or double doors used for admitting livestock to the side aisles. In most instances, the doors were single, about four feet wide, and with classic Dutch hinges. The older barns may have had such doors at either gable end; but in later times the rear doors were not common.

A few barns had double doors at the aisles of a total width of about eight feet. The Teller-Schermerhorn barn had double doors with beautiful wrought iron hinges. It may be that these double doors were in some manner related to the old-country break in the roof line.

In addition to Dutch hinges, the doors usually were fitted with an iron door pull both functional and decorative. Some doors were secured with iron latches decorated with chisel marks from a hot iron. Other doors had a sliding wooden latch operated with two pieces of rawhide hanging outside, one to secure it, the other to loosen it. Invariably, animal aisle doors swung outward, their pintles strong enough to serve for centuries.

Animal Aisle Doors.
Next pages. Left. Doors on the Teller-Shermerhorn barn and lean-to. The roof overhang and fascia boards are not original. Right. One of the animal doors in the Bratt-Mabee barn. The wrought iron hinges are of typical Dutch design.

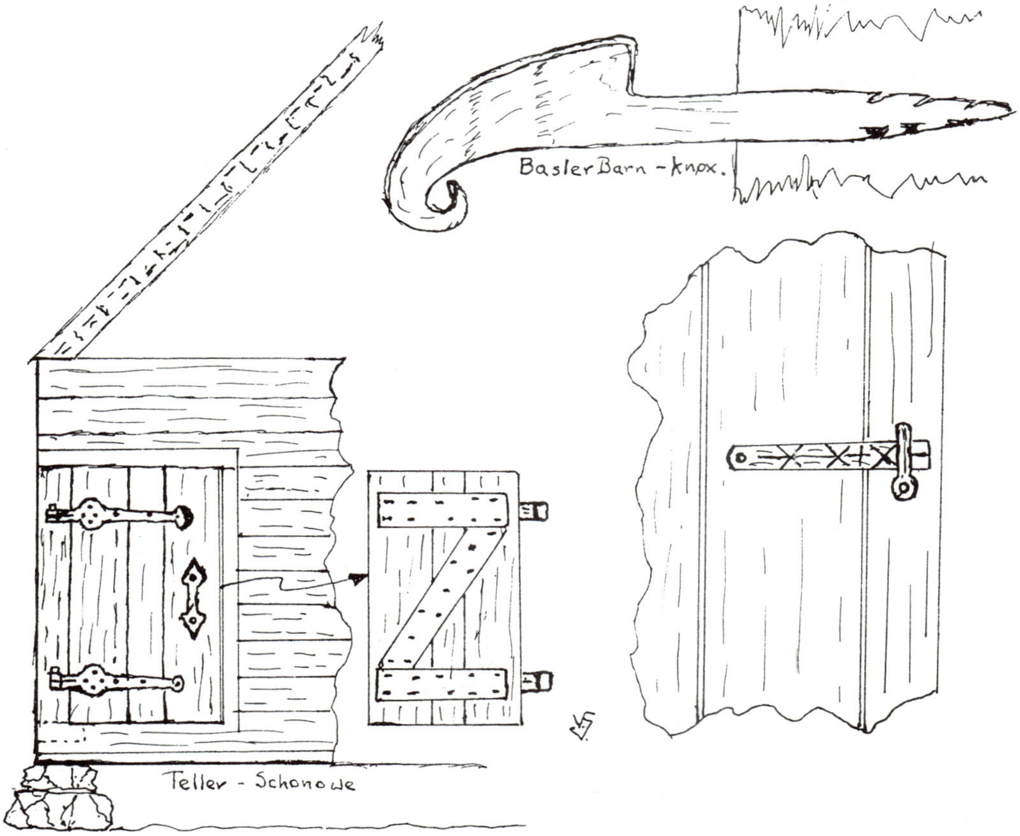

Wrought Iron Hinges

The homes and other buildings constructed in the settlements and surrounding farmsteads during the Dutch occupation of the Hudson, Mohawk, and Schoharie valleys have unique door hinges.

The earliest hinges (prior to 1750) apparently were made by master blacksmiths. Their elegant proportions and fabrication are a joy to see.

Unlike the strap and triangular-shaped hinges found on later structures, the Dutch hinge has its origins in the Netherlands and northern Europe. Its distinctive feature is an oval pad, located a few inches beyond the point at which the hinge is hung on the door frame. Beyond the pad, the hinge consists of a tapered portion whose length is about two or three times the diameter of the pad. It is fastened to the door with handforged rosehead nails, four in the circular portion, two or three in the tapered portion, and one in the flattened end. The oval pad on the hinge with its four nails provides great strength to the assembly.

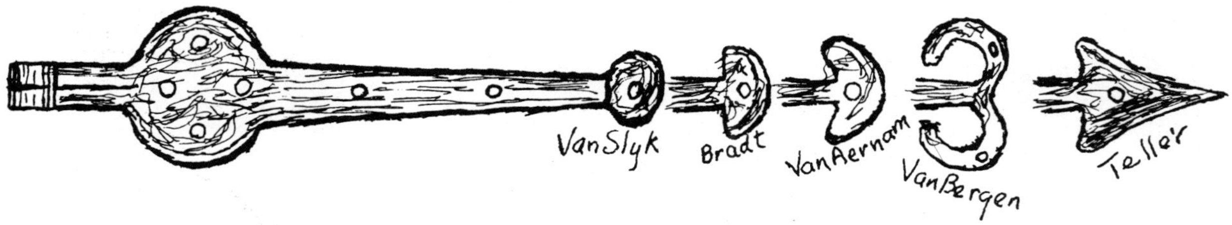

These hinges were used most commonly to suspend heavy divided Dutch doors in Dutch houses or the animal doors of Dutch barns. Four hinges were used for each divided door, and invariably they were made so uniformly as to appear to be cast from a single mold; but, of course, each was actually fabricated from a strip of wrought iron, probably at one central location, rather than by a local smithy.

The length of the Dutch hinge ranged from about one foot to nearly three feet. The largest hinges I have seen reportedly came from a New Jersey mill and are now found on the Bienstock-Johannes Decker barn off Red Mill Road, near Wallkill, N.Y., in the mid-Hudson Valley. The smallest I found were on the Schermerhorn barn at Schodack Landing and are somewhat similar to the two or three used in suspending the regular doors in the nearby house owned by the Rubensteins along River Road. The standard hinges found on the Dutch doors of houses and on the animal doors of barns have a length of 20 inches.

Most commonly, a single or double groove extends from the end of the hinge to the disk-shaped portion, so as to better fit on the suspending pin. At times, additional decorative grooves are cut at 45-degree angles to the long grooves. Occasionally, the outer end is split into two curlycues.

Occasionally, a door will have one Dutch hinge and a second hinge without the distinctive disc-shaped pad. In one case, on an ancient, solid wooden window shutter, the upper hinge had the circular pad, the lower one was without it, while the length, decorative grooves, and ends of both were identical. In this disposition of the two hinges, the upper hinge bore the greater load, while the lower one merely held the shutter in place.

An early smoke house with classic Dutch hinges.

Granaries

During its early development in our region, until the end of the 1700s, the Dutch barn was used primarily to store grains, especially wheat. The Mohawk and Schoharie valleys of upper-New York State were regarded as the breadbasket of the American Revolution, which is why Sir John Johnson and his allies swept through these valleys after the wheat harvest had been secured to burn the barns holding this precious supply.

The granary was made of wide, smooth, tongue-and-groove boards in an attempt to deter rodents. Many granaries still intact bear evidence that the integrity of the structure was often breached by rats and mice. Their holes were generally patched with zinc or tin.

Pentices

Most Dutch barns possessed pentices, stubby roofs extending outward from the gable ends above the wagon doors. They generally were wider by several feet than the wagon doors and extended outward some three to four feet.

These roofs were supported by light timbers mortised into the end anchor beams. There were generally three or four such mortises.

The pentice protected the sloping floor planks leading into the barn proper. Since the barns were invariably built on sills one to two feet above the ground, such a ramp was necessary to permit the wagons to climb to the level of the barn floor.

Roof Gutters

Gutters or eaves troughs of most early Dutch barns were made of a hollowed out timber extending the length of the barn. They were supported by brackets at a slope of 1 to 3 to ensure drainage.

Weather Vanes

Weather vanes often could be found atop one gable end of a Dutch barn, projecting above the roof line by several feet. On the Teller-Schermerhorn barn, for example, a metal weather vane was located at the south end of the steep roof. Unfortunately, it was blown away during the 1938 hurricane and was never found. Early photographs of the Wemple-Delamont barn show two vertical objects rising about two feet above the roof at either gable end. Whether they were weather vanes or other symbolic objects is not known, as both have been lost.

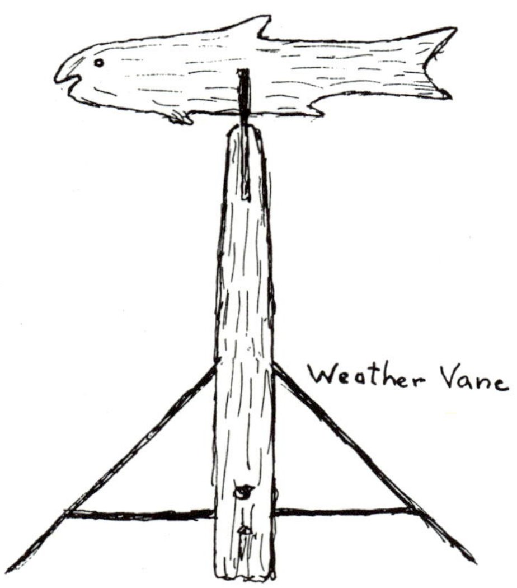

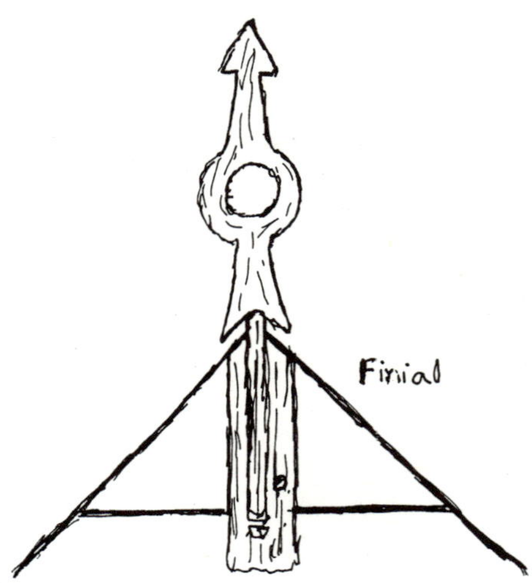

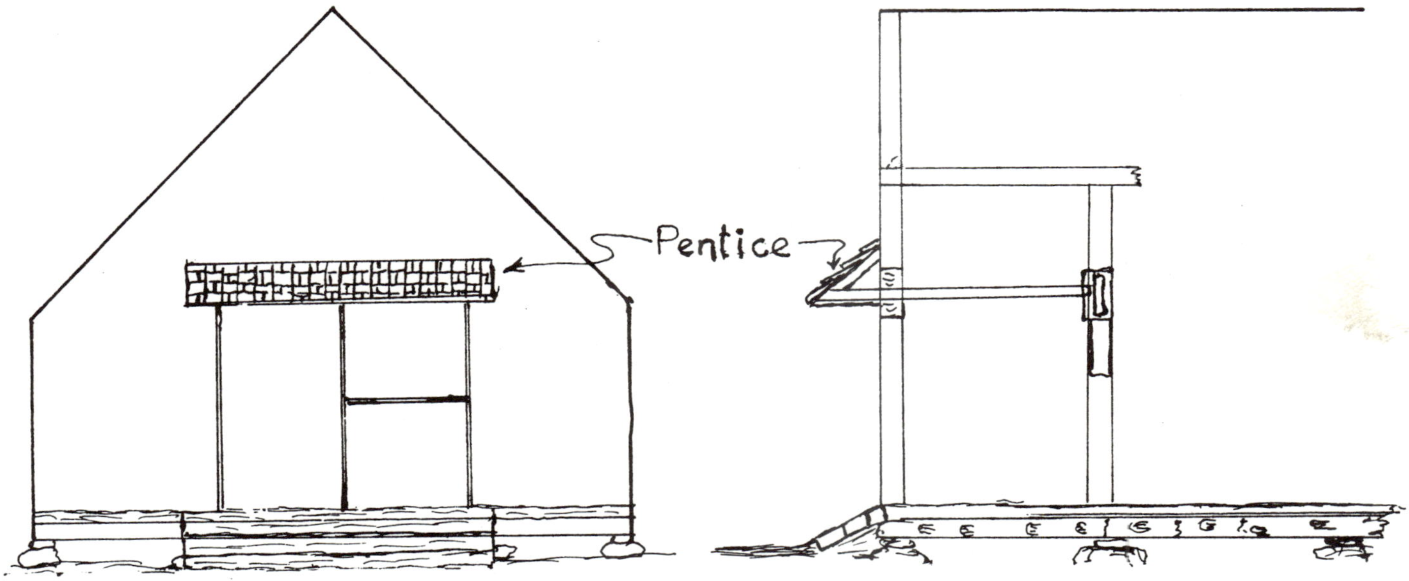

Teller-Shermerhorn barn showing pentice and weather-vane holder.

Latches

Among the smaller but more beautiful objects in the interior of a Dutch barn are the wooden latches and bars used to secure the granary and other doors of the structure. Most of them, as shown in the three sketches, are extremely simple, but functional.

The swinging latch sometimes is installed on the animal doors. The retainer of the latch complex consists of a notched round pin that projects through and beyond the inside of the timbered frame of the door and is held in place by a rectangular tapered wedge. The latch is retained by a single slotted bracket that limits its movement to an angle of about 20 degrees. This bracket can be mounted on either side of the outward swinging door.

A different sliding bar assembly is located on the inside of the animal door and retained by two brackets. It is usually manipulated by a projecting peg operated from the inside of the door.

Another simple latch, a miniature of that on the animal door, is found on the granary door. It is operated by sliding a bar held by two small wooden brackets.

Right: Top. Wooden latch from the Van Bergen barn. Middle. A typical small wooden latch. Bottom. Wrought iron latch from a granary door.

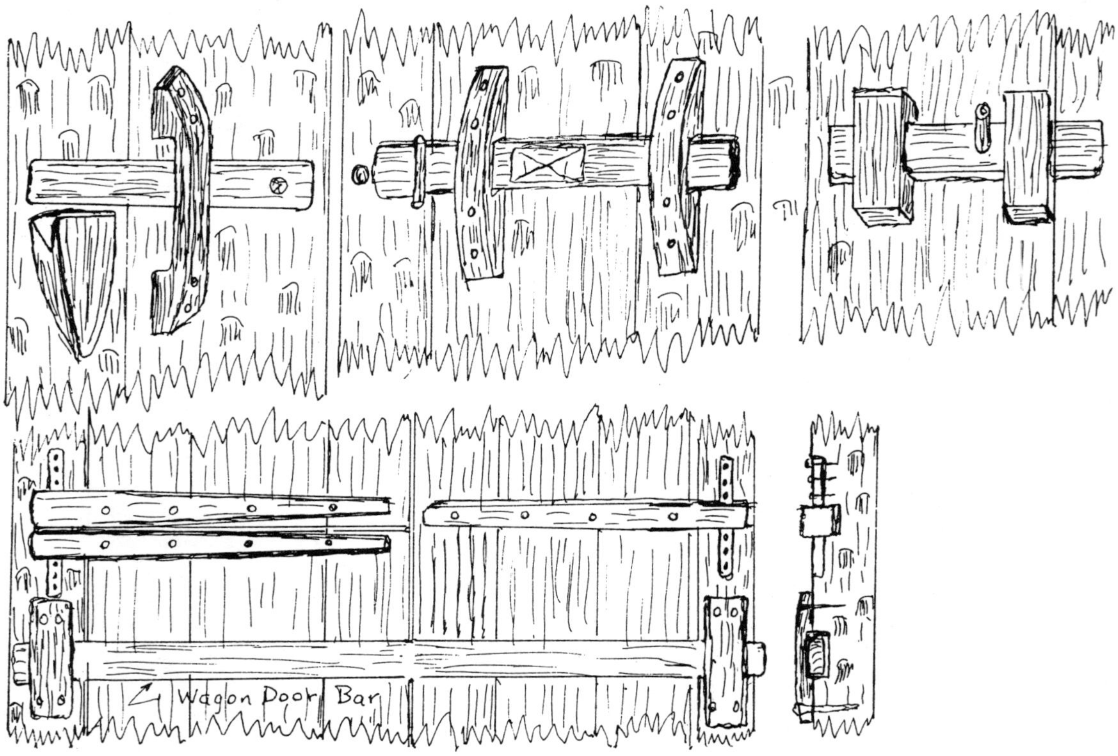

Ladders

In some Dutch barns, a built-in ladder permits access to the area above the anchor beams. Nearly every one of the pre-Revolutionary barns in the Town of Guilderland has such a ladder. All of them were built with rounded rungs, utilizing one of the vertical timbers as one of the sides. The other side was generally made of a smaller hewn timber.

A few similar ladders have been found in the Mohawk Valley in post-Revolutionary barns.

In the Teller-Schermerhorn barn, the ladder was built at the gable end, with flat rungs recessed in the vertical timbers supporting the outer wall.

Several free-standing ladders over 40 feet long and tapering from a two-foot-wide base have been found in Dutch barns. Their sides were made of a split tree, with its rounded sides facing out. It is difficult to visualize how such a ladder could be moved or manipulated.

Martin Holes

A distinctive feature of early Dutch barns served a double purpose: ventilation and access to the barn's interior for martins, swallows, and other insect-eating birds. The barn provided nesting areas and a rich source of food for such birds, who were particularly valued because they devoured mosquitoes.

The holes were large enough to permit the birds to zoom through without perching. Frequently there were three to seven of these holes cut into the wide siding at the gable ends.

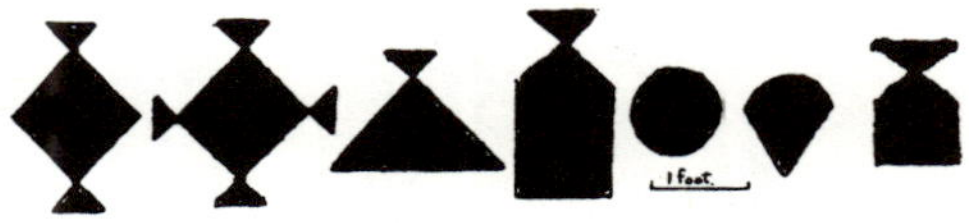

At least seven shapes have been observed in the outline of the Martin holes. Most were made square or rectangular, but almost invariably they were built with a secondary smaller hole, as shown in the sketches, illustrating the artistic sense of their builders.

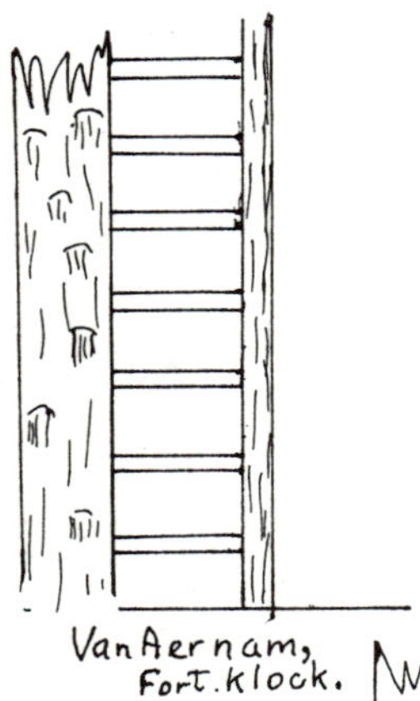

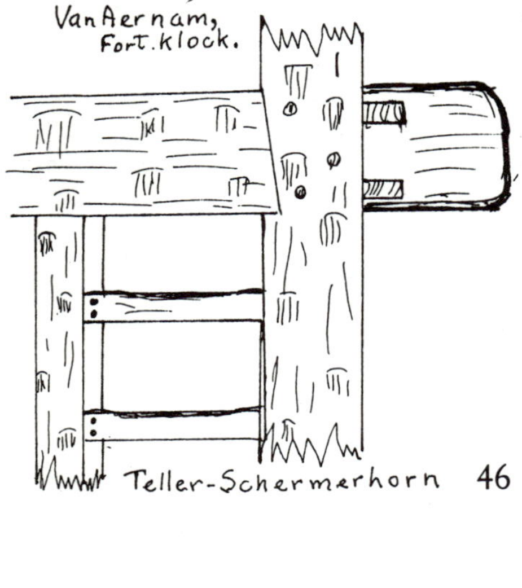

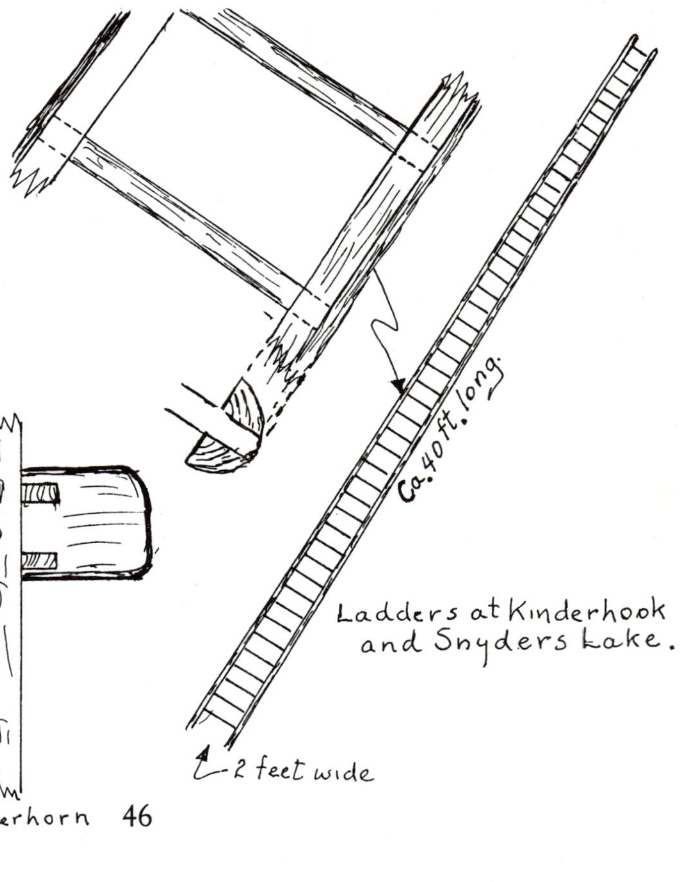

Loft Poles

A unique upper level floor was devised in Dutch barns for hay and grain storage. This consisted of peeled poles laid loosely on top of the anchor beams, extending to their furthest edges and spaced about a half-foot apart. This spacing provided ventilation for the hay or grain shocks above.

Some of the poles spanning the central area above the threshing floor would be removed temporarily so that the hay or grain stalks from the loaded wagon could be pitched up for storage.

Frequently, the timbers supporting the gable end pentices were made long enough to extend across the width of the first bay and were anchored in mortises cut into the side of the anchor beam. In such barns the poles supporting the hay were spread parallel to the gable end wall. In most instances these poles were made of straight trees from the forest, though occasionally hewn timbers were used. Sometimes remnants of earlier structures such as hay barrack roofs or poles can be found among the peeled poles supporting the crop.

Two of the massive anchor beams in the Bratt-Mabee barn at Rotterdam Junction. The anchor beam brace was the full width of the beam and had its edges chamfered. The loft poles, or hay-retaining supports, are clearly seen in this photo from circa 1940.

Dutch Barns of Schenectady County and Neighboring Counties

There once were probably many Dutch barns in the vicinity of Schenectady after it was settled by Arent Van Curler in 1661. The rich alluvial flood plains on either side of the Mohawk River upstream of Schenectady were highly favored by the Dutch farmer.

The interconnections between the early Dutch families of the Hudson-Mohawk region were very complex. In a number of instances, the death of a spouse was followed after a year or so by a second marriage. The families were extensive. Very frequently the children were named after their father, mother, aunt or uncle. As a result, it is often difficult to keep the relationships in proper sequence. If a son or daughter died at an early age, the next child born of the same sex frequently was given the name of the deceased. While primogeniture was followed to some degree, it apparently was not an inviolable rule, and gifts of land or other property were granted under many other conditions.

In general, such families as the Vedders, the Van Slycks, the Tellers, the Van Guyslings, the Wemps, the Bradts, the Vroomans, the Mebies, the Swarts, the Van Epps, the Glens, the Veeders, and the Schermerhorns were pioneer settlers who came from Northern Europe between 1620 and 1640 to Rensselaerswyck and Beverwyck. They joined Van Curler in forming the new village of Schenectady. They married girls between 18 and 25 years of age and established families on the farmsteads of the Great Flats or other flats adjoining the Mohawk River upstream of Schenectady.

The descendants of these first families established a very closely knit community through intermarriage. In the Massacre of 1690, some families were drastically disrupted and some families' lines virtually disappeared, but others survived and proliferated. The Vedder genealogy book lists more than a thousand descendants of Harmen Vedder.

The Great Flats on the south side of the river immediately upstream of the village were settled first followed by the smaller flats farther west. In Schenectady County, there were four major flats on the south side of the river, seven on the north side, and other lesser flat areas above the flood plain. The nine islands of the area were also available for growing crops, but none were conducive to the building of homesteads because of the annual spring flooding. In all instances, the farmhouses and barns were located immediately adjacent to the fertile flats, but were high enough above river level to avoid flood waters.

Over the 300 years since the first Dutch settlements, a great number of barns built in New York State have been destroyed, victims of fire, removal, or neglect. Most of the fires were due to spontaneous combustion from the storage of wet hay, although arson and carelessness have also claimed some barns. Removal of barns from one farm site to another occurred regularly up until about 1940, but rebuilding has generally destroyed the basic integrity of their structures. Therefore, Dutch barns have vanished from the region in increasing numbers. A few of those remaining have been restored as working barns or museums, and some have been converted to second homes for wealthy urbanites.

Neglect has also contributed to the disappearance of these noble structures. A leaky roof brings in moisture that quickly nurtures fungal spores that rot the large timbers. It is a sad thing to see the rapid deterioration that follows. Occasionally the powder post beetle will destroy barn timbers.

Recently, I asked six of my friends in the Dutch Barn Preservation Society how many Dutch barns are still standing in the Hudson, Mohawk, and Schoharie valleys. Their estimates varied from a high of 750 to a low of 300. This disparity makes clear that many new discoveries are likely to be made by the assiduous barn explorer.

Of the 14 Dutch barns I've known about in Schenectady County, only one remains. All were pre-Revolutionary period, and eleven of them were close to the Mohawk River.

In the pages that follow, I'll describe a sampling of my own discoveries of Dutch barns in New York.

The Van Eps Barn

When I photographed this 4-bay barn in 1938, it was in excellent condition. Located just east of Hoffman's Ferry on the north side of the road, it contained many early farm implements, such as wooden pitchforks and shovels. I was told some years ago that it had later burned.

The Van Eps barn on the right has been re-sided. The other buildings are later additions.

The deteriorating Van Bergen-Vedder barn.
Above. The sloped braces and cross ties, possibly not the original purlin plates and posts, may have been added when the roof was remade. Note the notches for rafters on both purlin plates.
Below. Several of the plank roof boards lean against the barn.

The Van Bergen-Vedder Barn

The Van Bergen-Vedder barn is said to date from 1680, making it the oldest barn in our region. It was built near the center of a large land grant obtained by Marten Gerritse Van Bergen. Located on a terrace above the rich alluvial flats bordering the Catskill Mountains, at Old Catskill near Leeds, not far from the village of Catskill, it contained several unique features. I first photographed this four-bay barn about 1950.

The anchor beams, made of pitch (yellow) pine and 32 feet long, had tenons which projected about 14 inches beyond the back surface of the anchor beam posts. They were 11 3/4 by 19 1/2 inches in cross section. The anchor beam braces were unique with their curved soffits reminiscent of those found in wooden ships. The full width of these braces attests to their early date. The age of the tree from which one of the anchor beams was fashioned was found to be in excess of 100 years.

One of the trustees of the Dutch Barn Preservation Society rescued an extended tenon of an anchor beam from the collapsed barn. The chamfer on the tenon is crudely made and much larger than later examples. From a pencil rubbing of the weathered wood at the outer edge of the tenon, we were able to establish the growth rate and number of annual rings of the tree from which the anchor beam was hewn.

Perhaps the most unusual feature in this barn was the presence of a second pair of purlin plates. These were supported by sets of slanting posts, each with a brace connected to a cross member considerably smaller than the main anchor beam below. The purlin plate as it was fastened to the outer edges of these slanting posts formed a complex of timber joinery identical in structure to the upper roof supports of dwellings in Holland dating back to the sixteenth century. I know of no other Dutch barn in the Northeast that has this feature.

Another unique feature of this barn was the Dutch hinge on one of the animal doors. The hinge's terminal end was splayed to form two half-circular endings. This was a thing of beauty.

Hugh (Sam) Van Orden, of Inbought Bay on the Hudson, frequently worked as a young man in the Van Bergen barn. He told me that his uncle, Henry Vedder, who farmed the land during the mid-1800s, learned that the floor of the barn consisted of oak trees split and laid side by side with their undersides resting on the earth below and their top surfaces adzed to flatness.

By 1980 local farming had ceased and the Van Bergen barn was left neglected. Its roof failed, so that water and rot invaded the timbers, and, within a relatively short time, the building collapsed. Good photographs were obtained of the ruined barn remains, so that a number of facts can be ascertained about its original structure: 1) It had a plank roof. The roof planks leaning on the western side of the collapsed roof clearly show the matching 30-degree pitch chamfers on their edges. 2) Both upper and lower purlin plates supported the roof rafters. I have a feeling that this originally produced a roof slope similar to that of the Teller-Schermerhorn barn. At some later date, the pitch of the roof was lessened, so the rafters rested entirely on the upper purlin plate. The support of the roof by the lower purlin plate was inadequate, forcing the outer walls to bear part of the roof load. Eventually, a wall collapsed, bringing the roof down with it. 3) Rotted rafters, purlin plates, and wall posts were the major causes of the barn's collapse.

The Teller-Shermerhorn barn. The author's house is left of the flagpole.

The Teller-Schermerhorn Barn

Once the finest of its kind in the Mohawk Valley, this Dutch barn was probably built by Johannes Teller on the edge of the Great Flats above Schenectady, in my opinion, possibly as early as 1701. Teller had been taken as prisoner to Canada by the French and Indians when they burned the outpost village of Schenectady in 1690. He returned from Canada a few years later and was given the Hindmost Farm No. 5 by his father, Willem, since he had no property of his own. About a hundred yards above the barn site, Johannes built a tiny one-room wooden house. Sometime between 1715 and 1720, he added to the house another room of the same design. Johannes died in 1725 and probably willed his farm, house, and barn to his first-born daughter, Margareta. Margareta, who married Jacob Reyerse Schermerhorn in 1712 at the age of 19, had seven children. When she died in 1741, the farm apparently went to her son, Simon Jacobse, born in 1730. Not until about 1760, years after Teller's death, did Simon J. Schermerhorn add on the much larger two-story brick house to the south end of the two connected houses that Teller had occupied for more than fifty years. Simon married Sarah Vrooman, and they had three children. Their second child, born in 1733 and named Jacob, married Engeltie Bratt. They had seven children, and, at the death of his father, Jacob took over Hindmost Farm No. 5. Their eldest child, Simon Jacobse, born in 1797, was living in the still-standing brick house in 1872, when he died at the age of 75 years.

This house may well have been built of bricks made from a clay deposit on the property. This is the house occupied by Clarence and Anna Delong Schermerhorn in 1933, when I bought a piece of the farm property from them.

When her husband died, I bought the barn on the Teller farm from Anna Schermerhorn. In 1952, I bought more land from Ann and Russell Horth, who had purchased the house and property from the Schermerhorn Estate. The Teller-Schermerhorn barn was unique among barns in the upper Hudson, Mohawk, and Schoharie valleys. Its roof, at an included angle of 70 degrees, was steeper and its sides, at 10 feet high, were lower than on other barns I knew, whose roof slopes ranged between 95 and 105 degrees and whose sides were 14 to 16 feet high.

One of the most interesting features of this barn was its double- sloped roof structure, achieved by raising the purlin plate. This double-sloped roof was similar to that found on barns in the Zeeland section of Holland and to that on the Decker-Bienstock barn west of Wallkill.

The original two-slope portion of the Teller-Schermerhorn barn resulted from the addition of an extension of the floor plan of the northwestern quadrant of the two bays, adding 250 square feet of flooring. To accommodate this extension, the side hill was excavated and a dry wall of Schenectady sandstone erected. This wall was 25 feet long, about 2 feet wide, and about 3 feet high. The wall sill and plate rested on a stone wall foundation. Evidence indicates the extension was added on after the initial barn construction.

When I dismantled the Teller-Schermerhorn barn in 1947--48, I discovered an intriguing feature whose role I did not at the time understand. I found that the roof rafters, massive hewn timbers 34 feet long, were in perfect balance where their notches fitted on the purlin plates. Each notch was only a third of the way from the thicker end of the rafter, but, as the rafter was tapered from top to bottom, the weight of wood above the notch exactly equaled the weight of the butt end. These rafters were fastened to the purlin plate with large wrought iron spikes.

The anchor beams, the posts connecting the anchor beams, the anchor beam braces, the purlin plates, and the roof rafters of the Teller-Schermerhorn barn were all made of local pitch pine. All of the barn's larger timbers, including the roof rafters, were shaped by broad axe and adze. The finished surfaces were beautifully smooth, as though they had been finished with a plane.

The siding was made of 1 by 10 uniform lapped white pine planks. The edges of the roof planks were beveled and overlapped in such a way as to shed water without leakage. Later the roof apparently was covered by shingles, which even more recently were covered by horizontal strips of tarred roofing sheets.

The barn sills were supported by massive stones mostly of limestone, granite, or gneiss. The former probably came from ledges along the Mohawk River above what is now Pattersonville, the latter two from boulders carried down from the Adirondacks by glacial action. The stones were piled under the load-carrying posts.

That there was at one time at least one hay barrack to the Teller barn is suggested by the presence of the metal holder of a hay barrack wooden screw. This device was used to raise and lower the movable roof of the hay barrack. Judging by the wear on its spike end, this wooden screw was used frequently. A more modern hay lifter probably was installed with new siding, as well as six double-louvered windows to ventilate the area above the top of the anchor beams. These were fashioned so as to shed rain but to admit air for drying the hay or grain.

The Teller barn was in active use until the end of World War II. Its last occupant was John Meyers, a horse dealer. He used the barn to shelter his horses and store hay and grain.

The Teller-Shermerhorn barn.

Left. Rear view. Mid-nineteenth century changes to the building included new siding, new roof overhang and trim, paired windows in the peak, shuttered openings for ventilation, and windows near the wagon doors (which hung on a track).

Above. The barn had both a finial (or lightning rod) and a weather vane.

Right. Interior view.

The Bratt-Mabee Barn

The Bratt-Mabee barn was built by the pioneer Arent Bratt family and was located on the Third Flat on the south side of the Mohawk River upstream from Schenectady. It was situated about a hundred yards from the river bank and at least twenty feet higher in elevation than normal river level, so only if a massive ice jam formed downstream would the barn be subjected to floodwaters.

This land was originally owned by Daniel Janse Van Antwerpen, an Indian trader born in Holland in 1635. He settled on the Third Flat about 1670 with permission from the Mohawk Indians and built a stone house which is still standing. The land was patented to him by Gov. Dongan in 1680. On January 22, 1705, he sold half, 63 acres, and the house to Jan Pieterse Mabee.

It is my feeling that the barn could have been constructed about 1712 as a gift to Caroline Mabee by her father Jan on the occasion of her marriage to Arent Bratt. The actual date of construction is unknown.

About 1735, Arent Samuelse Bratt built the Bradt House on the western edge of Hindmost Farm No. 1 of the Great Flats about 2 1/2 miles upriver from Schenectady. This farmstead adjoins Hindmost Farm No. 5, on which Johannes Teller built his house and barn. Bratt had built a barn next to his house, but that had disappeared by the 1920s, when I first traveled in this vicinity.

The Bratt-Mabee barn had all the features that characterize barns built in the early 1700s. Its massive anchor beams were supported by braces of the same width as the beam. Their surfaces were finished by adze, and mortises were carefully fashioned. The sway braces mortised onto the under surface of the purlin plate extended to the adjacent posts, to which they were mortised below the anchor beam, a sign of early construction. The original wagon doors at both gable ends swung inward on wooden hinges. Animal doors with classic Dutch hinges were located on both sides of both gable ends.

A unique feature on each of the anchor beam braces was a layer of thin boards nailed to cover two-thirds of the triangle made by the mortises between the anchor beam braces, the side of the vertical post, and the under surface of each anchor beam. Apparently these boards protected the nests of some fowl, perhaps guinea hens. A painted, egg-shaped pebble was discovered in one of these angles.

After more than two centuries and after a considerable portion of the lower south side of the roof had rotted out, the barn was dismantled by Mike Kelly and Don Carpentier. According to plan, it and two other Dutch barns are to be reconstructed near Fullers Station, Albany County. The barns are to serve as a corporate headquarters.

The Ellis Barn

This 3-bay barn was located on the Ed King farm, near the old Larabee Stop of the Fonda, Johnstown, and Gloversville trolley line connecting to Schenectady It disappeared in the 1960s.

The Van Slyck Barn

This was a beautiful 4-bay barn with low side walls and all of the other features of a very early barn. It was built south of the River Road close to the banks of the Plotterkill, but burned to the ground around the 1960s. Its owner, Andy Baan, gave me a number of beautiful Dutch hinges that he had salvaged from the ashes (see page 12).

The eighteenth-century Bratt-Mabee barn.
Located on the Third Flat on the south side of the Mohawk River, eight miles upstream from Schenectady. The siding has been replaced and the roof overhang and windows added.

The Brouwer Barn

This barn was said to have been the finest in the Mohawk Valley. It was sought out by representatives of Henry Ford, who wanted to move it to Greenfield Village at Dearborn, Michigan. The barn burned about two weeks after the visit of Ford's museum representatives.

The Viele Barn

This 4-bay barn was in perfect condition when I first saw it. It was located on the extreme western end of the Maalwyck, upstream of what is now Lock 8 on the north side of the Mohawk. Its large doors had wooden hinges, while the smaller animal doors had Dutch iron hinges with fine wooden latches. Unfortunately, this fine barn burned down about 30 years ago.

The Van Guysling Barn

I never saw the barn associated with the Van Guysling House, for it was probably taken down (destroyed) before the house was dismantled. The barn had been located on the western portion of Farm No. 1, land once allocated to Arent Van Curler, founder of Schenectady. The Van Guysling farmhouse had the steep roof lines of early Dutch architecture. It was still in fine condition when it was dismantled.

The Arent Bratt Barn

This barn was located west of the Bratt house (1735). I never saw more than its foundation stones.

The Wagner Barn

The Wagner barn, located near Poestenkill, is one of the longest and highest Dutch barns in the Hudson Valley. Although a late barn, it still retains most of the features of classic Dutch barns; but because of neglect to its sills and stone piers, it is rapidly falling apart. Its anchor beams are its only parts that still appear in good condition. Without piers and sills, the barn is actually cracking apart, a rare occurrence since, in most barns of this type, the core structure is the last feature to deteriorate, even if the walls go. At the peak of the barn, the sky can be seen where the connections have come apart.

A feature of this barn characteristic of several others is a shelf cut into the facing sides of two of the central anchor beams to accommodate an upper plank floor. The Wagner barn still has this floor intact, while the barns on the Stanley, Lord, and Herrington farms do not. A wagon door cut into the side of the barn permitted access to the second floor of this barn from an adjacent hillside, or, in the case of the Wagner barn, by a stone ramp. It is possible that the Wagner barn also was originally located along a hillside used to facilitate entrance.

The seven anchor beams of the Wagner barn are massive and have extremely wide tenons. They are of pitch pine, except for one of oak. Since the Herrington barn also has oak anchor beams, it is likely that large white oaks were present in the surrounding forest.

The Van Antwerp-Wemple Barn

This barn was originally located along Van Antwerp Road in Niskayuna. In 1987, it was partially dismantled and moved to another nearby location, where it was converted into a beautiful house. Many of the big timbers, anchor beams, and roof rafters were retained and exposed.

The Seeley Barn

This barn was on the north side of the Mohawk, downstream of the old Maxon Road Bridge spanning the river at the east end of Scotia. The barn is pictured in a news photo taken during the great spring flood of 1913.

The Larger Wemp Barn

I first visited the Larger Wemp barn more than fifty years ago and took some photographs of it. I was much impressed with the beauty and perfection of its construction. A second, smaller Dutch barn, known as the Smaller Wemp barn, stood not farm from it. At that time both were part of a working farm.

Jan Barentse Wemp had arrived in Beverwyck (Albany) about 1643. In 1654, he acquired property north of Albany, on the east side of the Hudson River, at the mouth of a substantial stream, the Poestenkill. Jan Wemp married Maritje Myndertse, and they had six children. The oldest, born in 1649, was named Myndert. He married Diewer Wendel, with whom he had two children, of whom the elder, Johannes (Jan), was born in 1670. In the Schenectady Massacre of 1690, Jan was taken captive to Canada. His father, Myndert, was killed.

Jan Wemp returned from Canada sometime after 1691 and, in 1701 he married Catalina Schermerhorn. They had four children before she died about 1708. In 1709, Jan was married again, this time to Ariaantje Swits. They had eight children between 1710 and 1726.

Jan Wemp was a skilled carpenter, and in 1711 he contracted to help build Fort Hunter at the mouth of the Schoharie River, west of Amsterdam. It is likely this was when he first viewed the 540 acres of land he was to buy from the Mohawk Indians in 1736. It is also likely he built the Larger Wemp barn shortly after he first settled in the region.

When the barn was dismantled in 1990, evidence indicated that it had been moved from its original site, quite possibly on the Mohawk River Flats, several hundred yards north of its later location. The move may have been to avoid the annual spring freshet, when the Schoharie ice breakup produced ice jams and severe flooding in that region. The date 1794 was found on the under surface of a sway brace, above an anchor beam.

The Larger Wemp barn near Fort Hunter on the Mohawk River. Two bays were added to the original four at an unknown date.

Photo by Clarke Blair

Dismantling, moving, and re-erecting the Larger Wemp barn.

Holes in the slate roof put the Larger Wemp barn in jeopardy. Fortunately, Carl Touhey agreed to purchase the barn in 1989 so that it could be re-erected on his estate. Work was started by the barn movers, Richard Babcock and his sons, Clayton and Charles, toward the end of March 1990. Despite snow, rain, and wind, the slate roof (not original) was salvaged, the roof rafters removed, and, in less than a month, all the timbers were disassembled. Much of this work was done with a gin pole, probably the same mechanism used in erecting the barn. During the month of May, a foundation and piers were prepared in a five-acre meadow close to the Touhey's historic stone home on Onesquethaw Creek, near the Albany County hamlet of Feura Bush, and erection of the barn proceeded with dispatch. By mid-June most of the barn was up. On October 13, 1990, it was dedicated with ceremony and festivities attended by a large crowd in conjunction with the annual meeting of the Dutch Barn Preservation Society.
Above. The interior a few months before it was dismantled.
Right. Disassembling the Larger Wemp barn.
Next Pages. Sorting the timber and raising the barn at the new site.

Above. Front (north) view of the Larger Wemp barn at its new site.
Below. Rear (south) view.

The Wemple-Diamond Barn

This was a very small two-bay barn with the largest anchor beam braces in the Mohawk-Schoharie area. It was neglected, and, by 1987, it was nothing but a heap of ruins.

The Wemple-Dellamont Barn

The measurements of the Bratt-Mabee barn were discovered to be basically identical to those of the Wemple-Dellamont barn in South Schenectady. It is quite likely that the same builder erected both barns.

The Wemple-Dellamont barn is one of the best of the remaining barns in the Mohawk, Hudson, and Schoharie valleys. Its main timbers, except for its floor and sills, are still in excellent condition. The metal roof which presently protects its interior is old but still intact, as are the barn's rafters, purlin plates, posts, and anchor beams of pitch pine.

I first photographed this barn about 1940, at which time it was still part of a working farm. Today, both wagon doors at the east and west gable ends are intact and hang on their original wooden hinges. The original animal doors remain with their wrought iron Dutch hinges. Much of the original barn siding remains.

A remarkable feature on the western wagon door is the rectangular wooden base of a large primitive lock, its keyhole still visible. The lock's presence indicates the value of the grain crop stored in the barn in its early days.

At the peak of the roof, on both gable ends, is a vertical reinforcement board, which held a finial, an ornamental detail. A photograph taken in the 1800s shows the barn to have had a weather vane on the western end of the roof.

Some time in the 1800s, a ramp was built on the south side of the barn which permitted wagons to be hauled to a temporary floor on top of the anchor beams. For protection against weather, a gabled dormer was built above the roadway. At a later date, the ramp was removed.

South of the barn, on a small terrace, is a gravesite with stones marking the burials of members of the Wemple, Dellamont, Van de Bogert, Vine, Pangburn, Van Valkenburg, Lewis, Miller, and Niver families.

The Normans Kill (named after Andries Bratt the Northman, a Norwegian) runs several hundred yards south of the barn, draining the hills back of Duanesburg with a fairly extensive watershed. About fifty years ago, an unusually high spring flood eroded a portion of the stream bed to disclose a crude woven wattle-work dam, perhaps a part of a primitive saw- or gristmill.

The Rau Barn

It is unusual to find Dutch barns very far away from major streams and their flood plains. Settles Hill, near Altamont, in the Town of Guilderland, Albany County, supports three exceptions, the Lainhart, the Rau, and the Stewart barns. All are late Dutch barns built when English barns were becoming the vogue.

The Rau Dutch barn has been in the Rau family since its construction, but only recently have its sills, flooring, and roof

Right. The Wemple-Dellamont barn. Original elements include the siding, pentice, doors, martin holes, and a finial at the peak of the roof. Later additions: the bermed wagon approach, plank bridge, dormer entrance, rectangular window near the peak, and attached shed.

been put into lasting order. It is owned by a very active member of the Dutch Barn Preservation Society, Everett Rau. The family barn is on a terrace in the so-called thermal zone, where drainage winds build up and nighttime outgoing radiation cools everything around the summit of the broad back hill.

This barn was probably built in the early 1800s as just a typical smallish Dutch barn, with anchor beams, braces, posts, purlin plates, and tapered rafters combining to form its characteristic core structure.

The Graulich Barn

In the rich dairy district between Cherry Valley and Canajoharie, along the south side of the Mohawk River near the small town of Argusville, a young farmer named David Graulich is putting together a brand new Dutch barn. When finished, it will be the largest and best example of this type of barn in existence. It will have a floor plan of 60 by 60 feet.

The massive timbers required to build such a barn were salvaged when a devastating tornado leveled a grove of virgin white pine near Hampton, Connecticut. David located a local sawyer, gave him a list of the number and size of the timbers needed, and now is in the process of fitting the timbers together. He hopes to have the barn erected by fall of 1993. With the help of his parents, he is making rapid progress with his project.

David plans to use his barn as a workshop. His basic design will follow that of the Larger Wemp barn, which was moved in 1990 from its original site near Fort Hunter, at the mouth of the Schoharie River, to the banks of the Onesquethaw Creek on the estate of Carl Touhey. The Graulich barn will be erected at the headwaters of Flat Creek, which enters the Mohawk River at Sprakers, New York. A number of excellent Dutch barns were located near Sharon and Argusville, all probably post-Revolutionary. One of the finest of these barns was dismantled and re-erected at Old Bethpage Village in western Long Island.

David's barn will include many innovations and will go a long way toward bringing recognition to the innate excellence of this type of structural design.

The Herrington Barn

One of the many fascinating aspects of the study of Dutch barns is the possibility of discovering previously unknown examples with their own unique features.

In March of 1991, a friend told me that he had seen the profile of what appeared to be a Dutch barn while traveling on a road southwest of the Tomhannock Reservoir in Rensselaer County. I drove out to see for myself. Near the heart of this region was a large assembly of very modern barn buildings. Beyond them, at the lower edge of a sloping hay field, was an ancient building that had the profile of a Dutch barn. A quick glance at its interior structure showed that it was indeed a three-bay Dutch barn with massive oak anchor beams. The two central anchor beams had shelves cut into their facing sides and other very unusual features. The shelves had apparently been cut into the anchor beams when they were first hewn. They were 2 inches wide, 12 inches deep, and extended nearly the full length of the 20 feet between the anchor beam posts. The posts, beams, and braces were all made of oak—a most unusual feature. Both central beams were adzed with a very wide and sharp tool. Other unusual features were the absence of wedges at the back side of the posts and inclusion of two martin holes shaped like equilateral triangles at each gable end of the barn. In other ways this barn was similar to other Dutch barns in our region.

Two more nearby barns have been found with similar shelves cut into the sides of a pair of facing anchor beams, the Henry Stanley barn and the Robert Wagner barn, both within a mile of the village of Poestenkill.

A House-Barn Complex in the Helderbergs

Frequently in Holland in the 1600s, barns and homes shared a single structure. In our region, a house-barn combination was located along the Beaverdam Road in the Helderbergs of Albany County. I made the photograph below in the late thirties, when the structure was already in a state of collapse. Unfortunately, I took only one picture, and close examination fails to provide conclusive proof that it had a Dutch barn core structure. There is a strong possibility, however, that it was fashioned this way, since the rafters visible through a hole in the roof and the four vertical struts at the gable end are similar to the typical framework of Dutch barns.

The chimney hole in the roof showed that the house section probably had a fireplace at the south end of the living space. The barn portion of the structure was long enough for four bays. It is not likely that there were any side aisles for animals. The pitch of the roof was steep enough to indicate that it could predate the Revolution. The barn siding remaining on the upper part of the gable end was of wide planks about 14 inches across.

Years ago, I saw a map printed in the middle of the nineteenth century indicating that the Helderberg Post Office had been at this site.

The simplicity and integrity in the design of Dutch barns persisted in the Hudson, Mohawk, and Schoharie Valleys for more than 300 years.

A possible house-Dutch barn complex in the Helderbergs.

Afterword

It is likely that the shift from a grain-based agriculture to dairy farming was the main factor in English barns gaining popularity over the Dutch. Another factor may have been the disappearance of large, tall trees in the woodlands surrounding local farms.

In the twentieth century, English barns in turn have been replaced by special metal sheds, which readily accommodate all of the sanitary requirements of the State Health Department for handling milking cows.

At present, the economic health of the farm establishments in New York State is in a serious decline. Small dairy farms have been so enmeshed in government rules and regulations that their future is in grave jeopardy. The Dutch barn is not readily adaptable to the installation of stanchions and the many other sanitary facilities now required. Hay no longer is stored exclusively in barns, but more frequently is packed into giant rolls that are so large they must be left in the field with little protection from the weather.

With more people desiring second homes in the northeastern New York countryside, Dutch barns are being acquired, dismantled, and reconstructed to serve such residences. In the process, most of the basic structure of the original Dutch barn is lost, even though its ancient timbers may be used to make a beautiful home.

In 1990, there are Dutch barns that have been preserved through the action of local historical and preservation societies, but some have been removed from their original locations and reconstructed. A few farm owners with a deep interest in preserving a link to the culture of the early Dutch settlers have protected their own barns in their original condition. Other barns known to me have been dismantled and are scheduled to be re-erected at sites where they will be preserved for posterity.

The fate of the hundred or so other Dutch barns in the area is very uncertain. With the rapid changes in ownership and the effects of economic stress on land values, it is likely that all of those barns not in active use on working farms will soon disappear, either through neglect, arson, or demolition by individuals interested in using the barns' massive timbers in the construction of expensive showplaces.

It is of the utmost importance that those barns still remaining on their original sites be measured and studied in detail, since they are the only reminders of the vital importance these structures once played in the economic and cultural development of the region.

***Left.** The Larger Wemp barn in the process of being re-erected on the Carl Touhey estate. The holes for pentice supports are clearly visible. Shown are Richard Babcock, barn contractor, the author, and onlooker Donald Sawyer. Photo by Clarke Blair.*

Glossary

Aisle Flooring
Floor planks in the animal aisles run parallel to the length of the barn (gable end to gable end). They are usually 3 to 6 inches thick and 14 to 20 inches wide.

Anchor Beam
A major horizontal timber running parallel to the gable end about 12 feet above the threshing floor. There are five in a four-bay barn.

Anchor Beam Brace
Braces in the oldest Dutch barns are as wide as their anchor beams and positioned at 35 to 40 degrees. Later ones are narrower.

Anchor Beam Post
The column supporting the anchor beam and, above it, the purlin plate which holds the rafters.

Animal Aisle Strut
A horizontal timber extending between pairs of bents at a height 6 to 7 feet above the floor.

Animal Doors
On either corner of the front of the barn. They are about 4 feet wide.

Animal Door Hinges
Traditional wrought iron Dutch hinges have a circular plate near the pintle and taper to a decorative, often arrow- or crescent-shaped, end.

Bays
A pair of bents fastened together with purlin plates, struts, and rafters. Most Dutch barns have four bays.

Bent
An anchor beam, pair of braces, and pair of anchor beam posts joined in an H-shape.

Cedar Shakes
Shingles made by splitting blocks of cedar with a mallet and froe and lapped to shed rain.

Clapboards
Exterior sheathing consisting of parallel boards with one edge thicker than the other.

Column Sill
A gable-to-gable timber which rests on stone piers to provide a stable foundation.

Eves Trough
A timber hollowed out and fastened below the roof's drip line to lead water away from the structure.

Floor Planks
Massive planks usually split from the center portion of large pine trees.

Finial
A decorative devise at the peak of the roof, often opposite a weather vane at the other peak.

Gable End Sill
A massive timber on stone piers that forms the base of the gable.

Gin Pole
A lifting device having a number of pairs of pulleys to provide mechanical advantage for raising heavy timbers.

Granary
A rodent-resistant storeroom for grains built in one corner.

Guinea Hen Roosts
Sides nailed on anchor beam braces to form nests for guinea hens.

Hay Barrack
A hay stack with a roof. Guided by 4 to 6 poles, the roof could be moved up or down using a lever or a screw.

H-Frame
A term used to describe the core structure which supports the roof of a Dutch barn.

Loft
The upper part of the barn in which hay, straw, and shocks of grain are stored.

Lower Transverse Strut
A horizontal timber between the outer wall post and anchor beam post.

Martin Holes
Martins, owls, and other birds had free access to the barn through these holes, which were often cut in distinctive patterns, in the gable-end sheathing near the top.

Median Floor Sill
A large timber resting on stone piers which runs the length of the barn.

Mortise
A rectangular socket cut in a timber, the female part of a mortise-and-tenon joint.

Mud Ramp
A sloping array of planks leading up to the entrance of the wagon door in the gable end.

Mud Sill
The supporting members on which the mud ramp is built.

Pentice
A small roof about 3 by 12 feet which protrudes from the gable end and protects the mud ramp from the rain.

People Door
A door adjacent to the wagon door.

Piers
Clusters of large stones which support the floor sills. In a 4-bay Dutch barn there are usually 25 piers.

Pintle
An iron pin on which the socket of hinge turns.

Plank Roof
An early roofing material made of 1 x 14-inch planks, 12-foot long, chamfered at 30 degrees.

Plinth
An alternative term for pier: stones resting on the ground which support sills.

Plumb Marks
Chisel marks which identified mortise-and-tenon joints and right and left sides of main timbers.

Purlin Plate
The timber held on top of the anchor beam posts which ran the length of the barn form gable end to gable end and supported the rafters.

Rafters
The long timbers which support the roof. Characteristically, they are tapered toward the top and balanced at the point they are supported by the purlin plate.

Raising Holes
One or two $1\frac{1}{2}$ inch holes bored through the anchor beam posts near their top. They served as sockets to support round rods used to hold a plank scaffold when the barn was erected.

Side Aisle Sill
Foundation timbers supporting the animal aisles.

Sills
The massive timbers which rested on stone piers and supported the barn above.

Sleepers
Another term for a Dutch barn sill.

Soffit
The undersurface of a timber.

Sway Braces
Supports which ran at an angle of 45 degrees or so from the sides of anchor beam posts to the soffits of purlin plates. They prevented the structure from swaying in a high wind.

Tenon
The projecting part of a timber which is fashioned for insertion into a corresponding hole in another timber. The male part of a mortise-and-tenon joint.

Threshing Floor
The central portion of the floor used to extract wheat or other grains by mechanical means, either with flails or the hoofs of horses.

Threshing Floor Bracket
A wooden bracket fastened to the side of a central anchor beam to accommodate a pole for leading a threshing horse.

Tongue-and-Fork Assembly
The joinery by which a pair of rafters is attached at the peak using trunnels.

Trunnel (Tree Nail)
A pin made of hickory or oak which was used to fasten timbers together in mortise-and-tenon joints.

Wagon Doors
The massive doors in the center at either end which swing inward and permit the farmer to drive through with a wagon load of shocks of grain or loose hay.

Wagon Door Posts
The framing which supports the wagon doors.

Wagon Door Hinges
These wooden hinges extended the full width of the wagon doors and swung on hickory pins sunk in the wagon door posts.

Wall Post
Supports the wall plate.

Wall Plate
Support for the base end of the rafters. It is the top of the side walls which are not load bearing.

Wall Sheathing
The covering of the sides and ends. Originally 1-inch thick planking, it was often replaced with clapboards or novelty siding.

Wedges
Tapered strips of hickory or oak used to hold the joinery snug.

Weather Vane
A devise on the peak to indicate the direction of the wind. On Dutch barns a fish motif was often used.

Appendix: A Dutch Barn Contract

In July 1993, a rare and important document relating to the Dutch barns of the Mohawk Valley came to light. It was an agreement between a farmer, Alexander Murray, and a builder, Abraham Smith, and concerned the construction of a Dutch barn in the Town of Princetown in Albany County. The document was donated to the Dutch Barn Preservation Society by Donald Keefer, Historian of the Town of Glenville and is presented here in transcription and facsimile.

An article of agreement made & entered into on this third day of December in the year of out Lord one thousand eight hundred and one between Alexander Murray of the town of Princetown and County of Albany farmer of the first part and Abrm. M. Smith of the City of Schenectady, the third ward of County aforesaid of the second part, Witnisseth that he the party of the second part for and in consideration of the sum hereafter mentioned does bargain, covenant and agree with him the party of the first part, to build & erect for him the said party of the first part a Dutch Barn of the following dimensions viz. forty two feet in length and forty in width—the stable outside the posts to be thirteen feet in height and the said party of the second part does by these presents firmly bind & obligate himself to hew the timber, having scnowers* found him by the said party of the first part and to set the frame having the raising expenses borne & then to finish & compleat the said building outside & inside in a workmanlike manner the materials all to be found and delivered on the spot by the said party of the first part After and in consideration of the said building erected & compleated as before specified by the said party of the first part does by these presents firmly bind & obligate himself to give and grant unto the said party of the second part the sum of seventy dollars, eight dollars when the timber is hewed, twenty dollars when the building is raised and the remainder when the building is compleated—he the said party of the second part does by these presents bind and obligate himself moreover to have the aforesaid barn finished & compleated on or before the first day of June in the year one thousand eight hundred & two in witness of our mutual agreement we do here unto set our hands the day & year before above written.

N.B. Dutch in line ten form the top intended before signed.

Alexander Murray

Thomas Bomayne

Abrm. M. Smith

* The meaning of "scnower" is obscure. From the context it appears likely that it was a word for a subcontractor whose trade was roughing out barn beams from the logs provided by a contractor.

An Article of Agreement made & entered into on this third day of December in the year of our Lord one thousand eight hundred and one between Alexander Murray of the Town of Princetown and County of Albany farmer of the first part and Abrm. M. Smith of the City of Schenectady, the third ward & County aforesaid of the second part, Witnesseth That he the said party of the second part for and in consideration of the sum hereafter mentioned does bargain, covenant and agree with him the said party of the first part, to build & erect for him the said party of the first part a Dutch Barn of the following demensions Viz. forty two feet in length and forty in width. The stable outside posts to be thirteen feet in heigth and he the said party of the second part does by these presents firmly bind & obligate himself to hew the timber, having scowers found him by the said party of the first part and to set the frame having the raising expenses borne & then to finish & compleat the said building outside & inside in a workmanlike manner the meterials all to be found and delivered on the spot by the said party of the first part. For and in consideration of the said building erected & compleated as before specified he the said party of the first part does by these presents firmly bind & obligate himself to give and grant unto the said party of the second part the sum of Seventy Dollars, eight Dollars when the timber is hewed, twenty dollars when the building is raised and the remainder when the building is compleated. he the said party of the second part does by these presents bind & obligate himself Moreover to have the aforesaid Barn finished & compleated on or before the first day of June in the year one thousand eight hundred & two in witness of this our mutual agreement we do hereunto set our hands the day & year before above written

N.B. Dutch in line ten from the top interlined before signed —

Thomas Romeyne

Alexander Murray

Abrm. M. Smith

Bibliography

For further reading, the most important text is by John Fitchen, a Professor of Fine Arts at Colgate University, and is entitled *The New World Dutch Barn: A Study of Its Characteristics, Its Structural System, and Its Probable Erectional Procedures.* It was published by Syracuse University Press in 1968 and is now out of print and scarce on the antiquarian market; although, it should be readily available in libraries.

For background reading, I would recommend *The Barn* by Eric Arthur and Dudley Whitney published by the New York Graphic Society (New York, 1972) and *New World Dutch Studies* by Roderick Blackburn and Nancy Kelley published by the Albany Institute of History and Art (Albany, 1987).

The *Dutch Barn Preservation Society Newsletter* has been published in spring and fall since 1988 and contains a wealth of important recent articles by society members. I was editor (and producer) of a randomly published journal called *Dutch Barn Research Miscellany* from 1988 to 1992. In all, eight issues were presented containing almost 200 subjects relevant to the study of Dutch barns.

The Dutch Barn Preservation Society

Like many other symbols of our agricultural past, Dutch barns are rapidly disappearing. Losses are resulting from the deterioration and collapse of abandoned barns, fires, the sale and removal of historic barns in whole or in parts for the construction of contemporary buildings and the outright demolition of these barns due to obsolescence or new development.

As a generation, we have the responsibility to see that these rare and significant artifacts from our agricultural heritage are conserved for future generations to observe, understand, and appreciate. When these barns can't be saved, it is incumbent upon us to record their presence through maps, photographs, measured drawings, and the compilation of historic records.

The Dutch Barn Preservation Society was organized in 1985 and incorporated in 1986 by the Regents of the State of New York as a not-for-profit organization by barn owners, historians, and other concerned individuals alarmed by the barns' rapid disappearance. Membership has grown steadily since the Society's inception, making new preservation initiatives possible.

The Dutch Barn Preservation Society sponsors recording projects, educational barn tours, and publications, and is a source of technical and historical information. Members have been instrumental in preserving and restoring threatened barns and are continually exploring opportunities to offer technical advice and financial assistance.
—The Dutch Barn Preservation Society,
PO Box 176, Rensselaer, New York 12144

Vincent J. Schaefer, 1906-1993

Vincent Schaefer left high school to help support his family as a machinist in Schenectady's General Electric Company in the 1920s. His reputation as an innovative model builder eventually landed him a position in the G. E. Research Laboratory where his scientific interests and expertise were developed while working as a research associate of Nobel laureate, Dr. Irving Langmuir. His many serendipitous discoveries, including weather modification through the principle of "cloud seeding," were the product of his unfettered curiosity, keen observation skills, and broad grasp of the inner workings of natural phenomena. He founded and directed the Atmospheric Sciences Research Center at the State University of New York at Albany in the 1960s and 1970s. He was awarded three honorary doctorates: Notre Dame University, York University, and Siena College. A lifelong outdoorsman, he formed the Mohawk Valley Hiking Club and planned the Long Path of New York. He was married in 1935 and raised three children in the "Dutch Colonial" house he built on a hill above the spectacular Teller-Schermerhorn Dutch barn. He was most comfortable in a wool shirt, khaki pants, hiking boots, knapsack on his back and camera in hand, ready for adventure.